Praise for *A Girl's Guide*

"This is the guide I wish I had before is a solid, factual, and practical guide to help young women make a major life decision with confidence. Amanda Huffman's thoughtful commentaries will provide readers with meaningful insights and positive motivation. Amanda tells readers, 'Know your value from day one.' This guide will make certain every young woman considering military service does just that and will be prepared to succeed, no matter what choice she makes. Strongly recommended."

—Mari K. Eder, Major General, US Army (Ret)

"... Amanda shares advice through her own personal experience as a former Air Force officer. ... Amanda weaves overall suggestions on what women should consider ... through the full lifecycle of a military career. ... This book is a perfect guide to help any woman considering life in uniform get straight talk on how it all works ... The best way to know whether the military is right for you is to pick up a book that tells you the ground truth from an eyewitness who wore the uniform of our nation. That book is Amanda Huffman's *A Girl's Guide to Military Service*."

—Jose Velazquez, Sergeant Major, US Army Public Affairs (Ret)

"... a how-to guide to join the military from a female service member's perspective, providing guidance and resources from inception through planning for the end of your military career. I liked that the author related her real-life experiences during her military career; it made it more relatable and easier to absorb all the knowledge and insight shared. ... a must-have resource for every mentorship program. On a scale of 1 to 10, I definitely give this book a 10! ... should be required reading for every young woman pondering the thought of a military lifestyle."

—Danielle De Leon Guerrero, US Air Force veteran, founder of Served Like HER, Inc.

"This book will change and guide many young women's lives for years to come. ... From detailing the all-important questions young women should ask themselves to decide if military service is right for them, to expertly explaining the personalities and missions of each service branch ... the critical difference between serving as a commissioned officer versus enlisting after high school ... the raw experience of the very-thorough medical examination at basic training, plus insight into the realities of workplace dynamics when you're the only woman (or one of a handful) in the unit ... Amanda's holistic view of women in military service, gained from her years listening to hundreds of women's service stories, across all service branches, and their military-to-civilian transition stories, enabled her to brilliantly structure this book.

It covers the most important topics to help young women decide which service is the best fit for them and encourages them to first understand why they want to serve our nation this way.

Amanda's wisdom shines brightly as she guides young women through the decision-making process, and for those who decide to serve, she adds many nuggets about the myriad of adventures, challenges, and experiences ahead. ...

Amanda reminds young women ... there's a powerful and active community of women veterans that make the military sisterhood a lifelong cherished treasure.

... a must-have resource for all high school counselors, vocational course teachers, parents and grandparents of young women planning their adult lives, daughters of immigrants (like me) with no family member veterans to guide them, college career centers, and public libraries."

—Graciela Tiscareño-Sato, US Air Force veteran,
author of *B.R.A.N.D. Before Your Resumé: Your Marketing Guide for Veterans & Military Service Members Entering Civilian Life*

"Wow! ... this book is incredibly informative and real! Transparency and authentic viewpoints are needed (and highly desired) when a young woman is making the decision to join the service. As a woman and veteran myself, so much of this informational narrative really spoke to me and what I wish I knew when I was deciding to join the military. ... this book really provides a streamlined approach that is easy to understand for all ages. It is clear, succinct, and doesn't make you read between the lines to get the information you're looking for! ... really great insights by a woman veteran, who experienced these things herself. This is not the kind of information a recruiter will share. I highly recommend this book to any young woman, and her family members, who are looking for real insights on considering joining the service. ..."

—Destinee Prete, PhD, US Army veteran,
president of We2AreVets

A **GIRL'S GUIDE** TO MILITARY SERVICE

SELECTING YOUR **SPECIALTY** ✶ **PREPARING** FOR **SUCCESS** ✶ **THRIVING** IN **MILITARY LIFE**

A **GIRL'S GUIDE** TO MILITARY SERVICE

AMANDA ✶ HUFFMAN

Elva Resa ✶ Saint Paul

A Girl's Guide to Military Service:
Selecting Your Specialty, Preparing for Success,
Thriving in Military Life ©2022 Amanda Huffman

All rights reserved. Except as excerpted in professional reviews, the contents of this book may not be reproduced by any means without prior written permission of Elva Resa Publishing.

All information and advice presented by the author is intended to inform and encourage young women interested in military service. Each situation is unique, and individuals should seek professional support as appropriate. There is no actual or implied Department of Defense endorsement. While all information was accurate at the time of publication, regulations change, so please consult with your recruiter, physician, or military leadership for current information.

Design by Sarah Flood-Baumann and Andermax Studios for Elva Resa, © Elva Resa Publishing.

Library of Congress Control Number: 2022943585
ISBNs 978-1-934617-67-0 (pb), 978-1-934617-76-2 (epub)

1L 2 3 4 5

Elva Resa Publishing
8362 Tamarack Vlg, Ste 119-106
St Paul, MN 55125

ElvaResa.com
MilitaryFamilyBooks.com

*Dedicated to all the women who have served.
Your stories continue to inspire me.
I'm proud to be part of the veteran sisterhood.*

*And to the next generation of women
who will take up the call to serve and
continue to break down barriers and
change the world.*

CONTENTS

Introduction   1

★ TO JOIN OR NOT TO JOIN   5

   1   Benefits   7
   2   Challenges   17
   3   Your Why   27

★★ FINDING YOUR FIT   33

   4   Military Branches   35
   5   Active Duty, National Guard, or Reserves   51
   6   Officer or Enlisted   55
   7   Career Fields   65
   8   Evaluations   73

★★★ BASIC TRAINING   79

   9   Mental Preparation   81
   10   Physical Fitness at Basic Training   89
   11   After Basic Training   99

★★★★ EMOTIONAL SUCCESS   101

   12   Transitions   103
   13   Stress Management   111
   14   Mental Toughness   117
   15   Self-Protection   123

★★★★★ PERSONAL SUCCESS   125
    16 Financial Choices   127
    17 Love and Relationships   135
    18 Motherhood   141

★★★★★★ CAREER SUCCESS   145
    19 Mentors   147
    20 Career Advancement and Promotion 153
    21 Exit Strategy   157

    Acknowledgments   163

INTRODUCTION

JOINING THE UNITED STATES MILITARY can be a daunting, yet rewarding, experience. Between acronyms and phrases you might not understand at first, what feels like far too many choices to make, and questions about whether the service is a good fit for you, the process can leave you feeling overwhelmed. This is especially true for young women, as I learned first-hand.

When I was considering military service, I had a vague idea that I wanted to join the military, but I did not know what it took to succeed or if I was even capable of serving. The doors to military service opened to me by chance rather than research. I knew I wanted to continue going to school and had been told by a recruiter that enlisting in the National Guard would give me the opportunity to serve while still attending college once I completed my initial basic training. I happened to have a friend in Reserve Officer Training Corps (ROTC), and when he heard I was looking to join the military, he took me to lunch and told me about how I could go to school and become an officer. That lunch changed my trajectory. I ended up joining ROTC and becoming a civil engineer officer in the Air Force.

I wrote this guide to military service specifically for women. While there are guides that cover the basics about the different branches, ways to serve, and preparing for military life, I wanted this book to offer deeper advice and feedback, from me and other women who have served, on what it is like to be a woman in the military. This book will help you make an informed decision by laying out options and covering the pros and cons, whether you are interested in officer or enlisted; Guard, Reserves, or active duty; Air Force, Army, Coast Guard, Marine Corps, Navy, or Space Force. It will help you discover the best path forward using your own goals, desires, strengths, and voice.

Many of the young women considering military service who stumble upon my podcast, *Women of the Military*, are searching for someone who can help them learn what it is like to be in the military and inspire them with their story. This book also includes personal stories and advice to support you on your journey. We'll start with essential information about branches in the military, ways to serve, preparing for your medical screening, and choosing your career field. Then we'll dive into keys to success as a woman in the military, with advice on how to mentally and physically prepare for basic training as well as tips for navigating specific challenges, such as sexual assault, motherhood, relationships, and post-traumatic stress. The passion I feel for helping young women in their journey of military service is written into each page. I want you to know you can do this—if you choose to—and I want to help you get the most out of your military career.

Deciding when to leave the military can be equally as challenging and important as deciding if you should join. Transitioning back from military service member to civilian again can present unexpected obstacles. I want you to join and leave the military with confidence in your decision. This book will build on that confidence to help you have a successful career, whether it takes the form of a four-year tour, serving until retirement, or anywhere in between.

This is the book I wish I had when I began to consider military service. I wish I had known that training is supposed to be hard, or that it is normal to feel scared and there are steps you can take to still go out with confidence. This book is in part the wisdom I gained about how much stronger I am than I realized at the time. I did things I did not think I could do. I gained confidence from my deployment to Afghanistan and learned the importance and value of the monotonous training we did over and over again until we could perform instantaneously when needed. There are the lessons I learned from my time in the service and the transition out of the military, along with the wisdom of so many other military women following their own unique journeys.

We are stronger when we support each other. Women have been

serving in the military since the Revolutionary War, when they dressed up as men so they could serve. With each new generation of military women, more and more opportunities open. The military as an institution still has its problems, and there are things that need to change. But I believe these changes are coming and I am hopeful for the future. I hope sharing some of the challenges we face today equips you to handle what lies ahead and build a better tomorrow.

My journey to military service began on the day the Twin Towers fell in New York City on September 11, 2001. My life has changed since then in ways I never imagined as a shy, timid senior in high school with undefined dreams. The military gave me countless opportunities and still opens doors as my career continues outside of military service. I pass this gift of experience on to you, sharing the tools you need to not only join the military but thrive within it and beyond.

TO JOIN OR NOT TO JOIN

THERE ARE MANY PROS AND CONS of serving in the military. Understanding the benefits and challenges of service is important for everyone. Even if you have already decided the military is right for you, you can use this information to help you throughout your career. Understanding the challenges of military service will leave you better prepared for the journey ahead.

If you are still considering military service, use the information in this section to consider what you hope to gain from military life. In the end, the choice to join the military is yours and yours alone. Use this section as a guide and then follow the path that works best for you and your life situation. The military changed my life in ways I never expected and helped me grow into the woman I am today. I believe there is a lot to gain from military service, and I am excited you are considering a path toward military life.

1
BENEFITS

THERE ARE MANY REASONS WHY WOMEN join the military, ranging from patriotism to opportunities for leadership, travel, and personal growth. Joining the military is a big decision. Choosing to defend your country carries many benefits, but it is also a significant commitment and not one that should be taken lightly.

When you join the military, you give up some control over many aspects of your life, such as choosing where you live. Military service also brings with it serious risks such as injury or death. Tragic events can occur in training accidents as well as wartime operations.

Creating a list of benefits alongside the challenges can help you gain clarity about your reasons for wanting to serve. Make two columns, one for the military benefits you find most desirable and a second for the challenges of military life that most concern you. As you read this book and identify benefits and challenges, note them in the columns so you can evaluate them later. Seeing the two columns side by side can help in the decision-making process as you consider whether or not the military is right for you. It can also help you make choices about military branches and jobs.

Many women find that the benefits they see in serving become their primary reasons to join the military or to reenlist or continue serving later in their career.

Serving Your Country

A common theme for veterans and service members is a love for their country and for serving others. Many are passionate about protecting personal freedoms and democratic values. Finding purpose and giving back to something greater than yourself helps you discover who you are and who you want to be. Don't discount the feeling of wanting to serve. Not everyone has that calling to serve, so if you do, lean into it and follow it.

Getting Out of Your Current Situation

Another common reason for deciding to join the military is wanting to get away from current life circumstances. For example, some people are from a small town and feel there are not many opportunities there. Others have difficult family backgrounds and see the military as a way to escape their past and start anew. Some dream of following a career path that requires either college or technical training, and the military offers the means to pay for their education. Whatever your current situation, the military may be an option to help you get a fresh start.

The People

The military gives members an opportunity to meet new people, and therefore, offers them a number of great opportunities. It starts from day one with the service member's basic training. Training puts a group of new recruits together in a unit and expects those members to become a team to help each other get through training. This team effort and regular connection with others continues as the members attend their technical training and get assigned to various bases around the world. The military pushes its members outside their comfort zone in many ways, one of the most important of which is meeting new people.

The military also creates a unique dynamic through its diversity. Service connects members together toward a shared mission, regardless of differences in gender, ethnicity, race, or sexual orientation. The military gives each unit a mission and they must find a way to work together to achieve that mission.

The bond that forms during the experience of serving extends into the veteran community too. Even after you leave military service, you will easily connect with others who served, and they will be more willing to help or find support resources because of your service. When I meet someone and learn they are a veteran, our conversation becomes that much richer because we both share an understanding of service and some of the experiences from it. If you are a veteran, you are part of a supportive community full of opportunity and connection. Millions of people across the country share a common bond because of their prior military service. The veteran community is thriving and active. Many local communities have networking and social events for veterans that can lead to fast friendships. The people you meet during military service are one of the best parts of serving, and many friendships from military life continue long past your military career.

Training and Career Options

Another benefit of joining the military is the training you receive. This can be leadership training or training in your career field. If you pick a career field you enjoy, you can use the skills you gain in the military to help you find employment in a similar field when you transition out of the military. Even if you don't continue in the same career field as your military career, there is plenty of training for both enlisted members and officers that can be translated to civilian jobs after the military or just in life in general.

Later, we will explore more about how to pick the right career field. For now, the biggest takeaway is that if you can dream it, it probably exists in the military. Throw out your preconceptions of what might be offered through military service and consider what you are most passionate about. Then start doing research to see what jobs involve those passions. Reach out to other veterans to help point you in the right direction. Depending on what you are most interested in, researching careers might also help you answer other questions about military life, like which branch to join.

Some job opportunities require a certain test score, degree, or being selected into a specialized program, so it is important to plan ahead and put in effort toward the career search process. Even with a lot of research and planning, you might find a job to not be a good fit. When this happens, the military offers its members the opportunity to train into new career fields. Your career is a journey, and you are not expected to have every answer ready from the start.

Regardless of the career each member chooses, compensation is based on rank and pay grade. This means once you leave for basic training or your first assignment, you already know how much pay to expect each month on the first and the fifteenth. A steady paycheck is certainly a benefit of military life, as many other fields do not offer that kind of consistency.

Structure

The military provides structure. There is a lot to gain in becoming more disciplined. Many who enlist after high school are able to thrive in a college environment after a tour in the military because they are a little older and have learned the skills of discipline and perseverance from their military service. If you are looking for structure, enlisting in the military can provide that.

The military allows you to discover more about yourself—a self you might not have known existed. Military service pushes you well beyond your comfort zone because even in training the trainers ask a lot of you. They push you to break you down and build you up. Most often your first assignment is away from family and friends. Being on your own lets you find out who you are and want to be.

The military also can be a place of comradery and teamwork. When you join the military, you become part of a bigger mission. Your training helps you and everyone around you understand there is no "I" in team. You find a way to work together through each step.

Travel

Another door the military can open is the opportunity to travel and see the world. Depending on your assignments you could end up living near where you grew up or on the other side of the globe. Different deployments, duty stations, and training programs provide a chance to see many parts of the US and other countries in a way most people never get a chance to. In my first year of being in the Air Force, I saw more of the US than I previously had in my whole life. In addition to moving and living in a new state for my first assignment, I also traveled numerous times for training.

My first international trip was my deployment to Afghanistan. While deployed, I was given a fifteen-day break during which the military offered to send me anywhere in the world. Most of my fellow military members went home to see their families. My husband (whom I met during ROTC and married right before I commissioned) and I decided to meet up for a romantic adventure in New Zealand.

That year changed my life as I saw the world in ways I never expected before military service. I lived, trained, or traveled through Kuwait, Kyrgyzstan, Afghanistan, Dubai, Australia, New Zealand, California, Arizona, New Mexico, Ohio, and Indiana. I made it almost around the world!

I also had the opportunity to turn a change of duty station into a road trip adventure. Some military members get the chance to live overseas in a foreign country and travel from there. The possibilities when it comes to traveling while serving in the military are amazing.

Leadership

Your branch, rank, and career field will determine the number of leadership opportunities you will have and the training you will receive. But even if you personally are not in a leadership role, there is still a lot to learn from both good and bad leaders within the military. The rank and order of the military makes it easy to determine who is in charge and, with the regular move cycle of leaders, most often you will get a chance to see a number of leaders. As you rise in rank, the military will begin to mold you into a leader. Both enlisted members and officers have the opportunity to lead.

I received numerous trainings focused on leadership throughout my officer training in college. My first six weeks as a brand-new second lieutenant in the Air Force were spent at officer training, being a leader, and making difficult choices. When I arrived at my base, I relied on the senior enlisted members (noncommissioned officers) to show me the ropes of military life. These enlisted leaders gave me a gentle push when helping me make tough choices, always giving me the final decision on how to move forward.

Education

One of the best and most well-known benefits of military service is the educational benefits you are eligible to receive. The Post-9/11 GI Bill helps cover the cost of tuition and books and provides a monthly stipend after you leave military service. But that is only one aspect of military education benefits.

The military also offers various programs that pay for you to attend college (civilian or military) and then, after graduating, you are sworn in as an officer into the military with a service commitment commonly between four to six years. There is also the option to use tuition assistance while serving on active duty to help pay for your degree while you are still serving.

There are a number of specialized programs the military uses as recruitment tools. If you are interested in being a doctor, lawyer, or dentist, or learning another specialized field, research or ask a recruiter about education benefits available to you.

Retirement

Another well-known benefit of military service is the pension. For a long time, if you served twenty years, you received a military pension, including health care, for the rest of your life starting right after you left military service. But if you did not serve twenty years, you received no benefits for retirement.

In January 2018, a new retirement system began. It changed the pension benefits that members receive at the end of twenty years of service, and also changed the retirement system for those who do not serve twenty years. The military now offers up to 5 percent matching into your Thrift Savings Program (TSP) account, a retirement savings and investment plan for members of the uniformed services, similar to a 401(k). It also includes a bonus at the twelve-year point as an incentive for members to continue serving. All those changes benefit everyone who serves, especially women. When you join the military, you can start investing in your TSP. If you decide to leave the military at any point, you will have already started to build up your retirement. Since women leave the military at a higher rate than men, this program has great benefits for women no matter how long they serve.

Health Care

Another important, but often not talked about, benefit of joining the military is free health care. And while the free health care provided by the military sometimes leaves something to be desired or a lot of red tape, it does have the advantage of giving its members peace of mind when it comes to not having to worry about medical expenses. Military members pay zero dollars out of pocket for medical coverage. The coverage for your dependents while you are serving on active duty can also be free or there might be a minimal fee that allows more choices when picking a doctor. If the military is your first job, you may not truly understand the value of this benefit. But with the rising costs of health insurance and the premiums people with civilian jobs pay, this benefit is an important one to consider when looking into military service.

Positive Changes for Women

Being a woman in the military is a challenge, but the culture of the military is changing positively toward women.

The controversy about women being in combat roles has a long history. First, all military roles were filled by men. Women were not able to serve in the military. Over time, the military began allowing women to serve in limited roles, such as nurses and secretaries. But soon women were serving in all kinds of roles. With the onset of World War II, women were given more opportunities to support the war effort. While women were still the minority, their role continued to grow.

In 1993, women were given the opportunity to serve in combat roles that had previously been off-limits to them. Fighter pilot roles in all branches and combat vessels in the Navy and Coast Guard opened up to women. During the wars in Iraq and Afghanistan, women filled roles such as mechanics, medics, engineers, and police officers. With these jobs, women found themselves in combat, even while combat exclusion existed. As a result, many of these women were not given the medals they earned—not because they hadn't seen combat, but because they weren't "supposed to" be there.

By the time I deployed in 2010, women were playing an active role in helping the military connect with the local population. Women were seen as an asset on the battlefield (even when they were not being recognized for their role). The military saw the need for women and began quietly testing Female Engagement Teams (FETs) that were attached to infantry, SEAL teams, and other elite positions in which women could not officially serve. Through these experiments, combat exclusion was eventually repealed in 2013.

In 2016, the US opened all career fields for women. It was a long process to ensure jobs were not restricted based on gender. It took many women stepping up to serve in non-traditional roles and not only proving they could do it, but showing they were a needed asset for the mission.

The military also began changing the maternity leave options for women. Beginning in 2016, women in all branches could receive twelve weeks of maternity leave. In 2021, legislation was introduced to increase Department of Defense parental leave provisions for military members and secondary caregivers. In addition, new moms do not have to meet physical fitness standards until twelve months after giving birth. Now, mothers have a year after giving birth before they can be required to deploy overseas. The military is continuing to address other ways to be a better place for women.

Today's military gives you the opportunity to serve in any career field, regardless of gender. Your role is based on the work you can do and how you can help the military meet its mission.

Sense of Purpose and Personal Growth

The military can provide a sense of purpose in a way few other organizations can. This purpose drives service members to go through hard experiences and grow into stronger people. The military sends service members around the world to fight wars, provide humanitarian aid, and staff bases overseas. These unique experiences can open your worldview and change your path forward.

Each person's military journey is unique. You can't predict yours until you begin and it will likely change as you go through your military career. Be open to the possibilities! If you decide the military is the right place for you and you answer the call to serve, know that your life will never be the same.

There are plenty of other benefits to joining the military that only come through experiencing military service. This isn't an all-inclusive list of benefits or reasons people join the military. Think of this chapter as the beginning of a brainstorming session to get you thinking and help build your pros and cons lists.

If you haven't started creating your list of benefits yet, pause now and do it. You can make your list on your phone, a piece of paper, or whatever way works best for you. This is a great way to get your thoughts out of your head as you decide if the military is a good fit. You can look back on your list as you move forward with or without the military.

2
CHALLENGES

JUST AS THERE ARE BENEFITS to serving in the military, there are also challenges. The military demands a lot from its members. You will have to meet service requirements, spend significant time away from loved ones during deployments, and live wherever the military sends you. These and many more challenges are some of the reasons women decide not to join the military or decide not to make a career of it after their initial service commitment.

Disqualifications

If you do not meet the minimum requirements to serve, then your military career will end no matter how you feel about military service. One of the most common disqualifications is medical. The military has strict requirements to be medically cleared. The goal of the Military Entrance Processing Station (MEPS) is to weed out those the military deems medically not fit to serve. Potential service members go through a rigorous physical exam and medical history review. Common disqualifications include asthma, major allergic reactions, drug use, and not meeting height and weight standards.

Waivers are possible for some medical issues. It is important to be aware of what can disqualify you from military service so you know from the beginning if you will be qualified to serve. In chapter eight we will dive into what the MEPS physical experience is like.

The military also reviews your criminal record. Having a felony will require a waiver and could lead to a disqualification.

Another challenge you may run into when joining the military is not having the education requirements or not achieving a high enough Armed Services Vocational Aptitude Battery (ASVAB) score for the career you desire or to serve in the branch you want to join. The military limits the number of members who have a GED, so if you do not have a high school diploma, make sure your recruiter knows so they can help file the correct paperwork to help you bypass this complication.

In addition to your academic credentials, your score on the ASVAB determines if you meet the military's requirements and can also determine what jobs are available for you to consider. There are different requirements across branches, between enlisted and officer, and between career fields. The higher your score, the more opportunities will be available. One thing to note is that there are limits to how often and when you can retake the test (one calendar month after your initial test, another month to retest a second time, and after that another six months before you can retest). Your final score will be the score you receive on your most recent test. Make sure to study and prepare for the ASVAB. Check for current information at the official ASVAB site, OfficialASVAB.com.

Deployment

One of the biggest things to consider is the possibility of deploying, including to a combat zone. The role of the military is to protect its country. You need to be prepared to deploy overseas for both humanitarian and combat roles. Sometimes a deployment could happen in the country you are serving. It does not matter if you serve on active duty or in the National Guard or Reserves. All military service comes with the possibility of deployment. National

Guard members deploy overseas and can also be activated for state or federal response. Serving in the military means you understand the sacrifice required of service. Military members must be willing to give their life for their country and do whatever the military asks of them. While serving in the Air Force, I deployed with the Army on a combat mission. It was not what I expected when I signed up to serve in the Air Force but that did not change the requirement of service. If you are joining the military, prepare to deploy overseas, even if it never happens.

Do not join the military expecting that deployment will not happen because the country you are serving is not at war. Things within the military can change in an instant and there is so much more going on in the world than what ends up on mainstream news.

Separation From Loved Ones

In the military you will often find yourself separated from your family. Basic training can last from six weeks to three months. Career training school typically takes two to six months, but could be longer or shorter depending on the career field.

If you decide to serve on active duty, your first assignment could be anywhere. You may end up relying on video calls and emails to stay in touch with family back home. Missed holidays, weddings, and other significant events will also happen because your time off (called *leave*) might not be approved or the cost to travel might be too high. Missing those first few holidays on active duty was a hard transition for me. But I have been able to make my own traditions and cherish the times I do get to be home with family.

Even if you get married and start your own family, you can still find yourself separated from them. Trips for the military are so common they have their own acronym. Temporary Duty or TDYs are military work trips. They can include training in the field, a business trip, a certification course, and more. TDYs can last a few days to a few months depending on the purpose of the assignment.

Military couples and families spend more time apart than most American couples. My friend was in a doctoral program and asked her civilian classmates how many of them had to spend more than a week without their spouse. Very few hands in the class went up. And when she asked if anyone had spent a month apart from their spouse after they got married, no hands went up. There would be a totally different response to that question in a room full of military spouses. Most military spouses have experienced periods of time apart from their service member due to military service. My husband and I have spent many nights apart due to military commitments. It is challenging. But it is not impossible to have a strong relationship with your significant other and children even with all the separations.

Deterrence of Family and Friends

Being deterred to join the military by family and friends is particularly common among women considering military service. The military is often seen as a "man's profession" and some family or friends may want to "protect" their wives, daughters, sisters, or friends from joining the military. The wishes of well-meaning friends and family is not a good reason to not serve in the military. They may want to protect you but ultimately it is your life, and you have to live with the choice of service. Deciding if you want to serve rests solely on you.

It can be a good idea to reach out to others to learn more about the military. The military is not right for everyone and it's wise to be informed before you make such an important decision. In addition to resources like this book, you can also go to the local recruiter's office and talk to a recruiter. They are there to help answer your questions. You do not need to know what branch you might want to join. A number of recruiters I have met over the years will help recruits honestly assess if the military is a good fit for them and direct them to a different branch if they think that might be a better opportunity. Be open about the main reasons you are considering military service.

LinkedIn is another great resource. Many military veterans put their service experience in their profile. Through a quick search you may be able to find women veterans in your area that you can network with. These are people who can share their own experiences serving, including how they may have dealt with people trying to deter them from serving because they were women.

If reaching out to someone you don't know feels too intimidating, you might already know someone who has served. Share within your social circles that you are considering joining the military and see what connections you have. You might be surprised by the resources already at your disposal that you didn't know existed.

If it feels like you do not have anyone to talk to and you are not ready to talk to a recruiter, you can always reach out to me. I am happy to connect young women joining the military with veterans and service women who can help answer questions.

In the past I used to say that you should join even if your significant other is deterring you from joining. And while that is the general advice I give, Lizann Lightfoot, author of *Open When: Letters of Encouragement for Military Spouses,* shares a different perspective. When her boyfriend (now husband) joined the military, the recruiter said not to listen to her because the choice did not affect her life. But in truth, it had a huge impact on her life. And not just when they got married but from the moment he left for basic training. Looking back, she wished she had been included in the discussions when her boyfriend was joining. It may not have changed the outcome and the final choice would have been up to him. But being part of the discussion and knowing the impact it would have on both their lives would have made the experience easier. If you are in a serious relationship and your significant other does not want you to join the military, you should discuss what you both want for the future. If you decide to join the military, your partner will undoubtedly have to make sacrifices along with you.

Talk with your significant other and spend time doing your own personal reflection. When you think about the future, will you regret not serving? Will you have resentment over this if your significant other stops you from serving? While relationships are important, doing what is best for you is what is most important. Having to live with regret isn't something you want to do. Even if your significant other is at first not supportive of your choice, that does not mean the relationship has to end or that you should not join the military. It just means more discussion needs to take place. The military will affect more than just your own life.

Not Having Control Over Where You Live

Sometimes the lack of control service members have over where they live stops people from joining the military. Most active duty service members move every two to three years. Sometimes a service member can submit requests or preferences for assignment locations, but the military will issue orders based on where you are needed.

If you want to have more control over where you live, the National Guard and Reserves most often allow you to choose where you live for your home station. I will share more about the National Guard and Reserves in chapter five. Right now, just know it is an option and could give you the ability to serve without having to move.

No matter how you serve (active duty, National Guard, Reserves) the military will require you to spend time away from your family for basic training, trainings, and other service requirements. If you are interested in living in a particular area and do not want to spend extended time away from your family, then joining the military is likely not a good fit for you.

Service Commitment

Another consideration when joining the military is the level of commitment. The military is not like a civilian job where you can put in your two weeks' notice and quit if you don't like it. The military often has its members sign a four- to six-year contract for service. It can be even longer depending upon the education and training required for the chosen career field.

Joining the military for four years might feel like a long time, but in the grand scheme of things the military commitment is a few short years of your life and can lead to many opportunities.

Recruiter Experience

Sometimes people decide not to join the military because they do not have a good experience with their recruiter. Recruiters work for the US government and will sometimes put the needs of the government before the personal interests of the person they are recruiting. When they are busy recruiting people for their branch, they might not make time to help determine if a different career field is a better fit. It is easier for them to fill open spots that are in demand than for career fields that have a long waitlist to join. Also, some potential recruits get cold feet in the months waiting to join, so pushing them toward open positions can help ensure new recruits decide to join the military.

Finding the right career field is one of the most important aspects of joining the military. Do not let a recruitment bonus determine your career path. Make sure if there are bonuses available (and if that's important to you), that you do your research about those career fields and pick the best one for you. You also do not have to join a career field that has a bonus attached to it. Bonus or no bonus, figure out what you want to do and then follow that path. You will be happier in a career field you love, even if it means missing out on a bonus.

Another consideration is your options for enlisting or becoming an officer. If you are in a recruiter's office feeling ready to join now, the longer process to join the military as an officer may influence your decision. But it is important to consider all your options and not just follow a recruiter's opinion that enlisting is the best path for you. If you already have a degree, take a look at the opportunity to become an officer to see if it's a good option for you. You can also choose to enlist and become an officer later, and there are positive aspects of going that route.

When I was looking to enlist with the ultimate goal of getting my college paid for, the recruiter never mentioned the option of becoming an officer through the ROTC program. Fortunately, I did find out about this way to continue to go to school while working to become an officer in the military, and that became my path.

One of my friends was struggling through college and was ready for something new. She decided to enlist in the Air Force instead of continuing with her education. For her, enlisting was the right path. And then when she got out of the military, she was able to use her education benefits and graduate.

The truth is recruiters do not and cannot know everything. They only know what they have gained from their own experience and training. This book is a resource but even it cannot answer all the questions you may have. Use the recruiter and any other resource as a tool and then do your own research and reach out to others.

Male-Dominated Environment

Another important aspect to discuss when considering military service is the male-dominated work environment. The percentage of women in the military is on the rise, but is still very much a minority in all branches, with the Marine Corps having the lowest percentage of women. It isn't easy being the only woman or one of a few women in your workspace.

But the male-dominated workforce has much bigger implications than the loneliness women often express when serving in the military. Sexual harassment, assault, and rape are systemic problems within the military. I experienced varying levels of sexual harassment in the workplace. That does not mean it will be your experience, but it does mean it is important to be aware of this issue so you can better protect yourself.

The Department of Defense is working to root out the causes of this negative culture and change the process for how cases dealing with military sexual trauma (MST) are addressed. Later in the book, we will talk more about what MST is, tips on how to protect yourself, and how to report it if it happens.

IS THE MILITARY RIGHT FOR YOU?

After considering the negative aspects of military service and the lifestyle required, some people may decide the military is not worth the sacrifice or that it doesn't feel like a good fit.

It is important to figure out if you are willing to move away from family, travel for various trainings, deploy overseas, follow orders, and more. The military asks a lot of each member. And there is not a turning back point when the military asks you to sacrifice more than you expected. Military members are known to miss important family events such as weddings of immediate family members, anniversaries, holidays, and birthdays. The needs of the military come first, and it is important to understand that when joining and to determine if you are ready to make that sacrifice. Some people end up being disqualified for service and the choice of whether to serve is made for them. Others walk away from military service because the sacrifice required does not line up with where they want to go.

While this chapter covers some of the most common reasons people decide not to join the military, there may be other challenges (or benefits) that are important to you. If any of these things make military life not feel like the right path for you, dive deep into those feelings. Understand what is stopping you from joining and then do your research around each topic. If you are afraid of moving, going to war, or even attending basic training, know that is completely normal. Being afraid or having apprehension is not a reason not to join. Moving away from family and friends is difficult, but the new friends you meet and the places you get to see make that pain less challenging.

I was afraid to deploy. Going to war is scary. Thinking about the possibility of not coming home and leaving your family to go on without you is one of the hardest parts of leaving for an overseas deployment. But deployments are also an opportunity to grow and see the world. My deployment pushed me outside my comfort zone and led me to accomplish many things I never expected I could do. My fear of the unknown was the biggest thing holding me back. The military did not give me a choice about it but my training had prepared me. They told me what to do and I followed orders. And I learned I could do way more than I ever expected. That experience changed me into a stronger person.

My whole military experience could be categorized as facing

challenges and then being pushed outside my comfort zone to overcome them. The military felt like an overwhelming challenge. But I decided to take a risk and join, and it changed me into who I am today. I feel so lucky to have had the opportunity to serve.

Is the military right for you? Take a moment to reflect on your thoughts about the benefits and challenges. Update your pros and cons list. What else do you want to know more about? What is pulling you toward military service?

3
YOUR WHY

WHY DO YOU WANT TO JOIN THE MILITARY? Knowing "your why" will help you make decisions about your future with confidence and ground you later during challenging times.

Joining the military is a big responsibility and each person in the military plays a key role in keeping our nation safe. We answer the call to serve and make up the military in all different roles, branches, and jobs. Without an all-volunteer force, the military would not exist as we know it today. But what drives people to the military is unique and often depends on their life circumstances or goals for the future. When you know why you want to serve and what draws you to military service, you can use the information to help you determine if the military is right for you and help you as you make every other choice about your service.

Recalling your why can encourage you in hard times. For me, my why helped me when I was ready to quit basic training. Basic training is not easy. And it may feel easier to quit and not have to run or get yelled at anymore. I did not allow myself to consider the notion of quitting because I had my why of getting my college paid for and serving as an Air Force officer. I had already devoted two years of my life in ROTC to get to that point, and I knew that somehow I would get through the stress of officer training.

Was it hard? Yes.

Were there days I didn't think I could do it? Yes.

But I knew what I wanted and I focused on that goal. It drove me to not give up and to keep taking it one day (and sometimes even one obstacle) at a time. My why helped me in those early days when my confidence waned.

HOW YOUR WHY CAN AFFECT YOUR CHOICES

Why you want to join the military says a lot about who you are as a person, and it might be something you can use when determining what branch of the military you want to serve in. Or maybe it can help you determine your career field. It can even help you determine when it is time to leave the military.

For example, if you are joining the military because you are looking for a way to get your education paid for, knowing about tuition assistance can help you begin using the education programs offered to military members from the beginning of your career. It may be possible to use tuition assistance and to have a concurrent commitment requirement that will not require you to extend your time in service. This gives you the option to jump-start your degree while serving and finish your degree with your GI Bill.

Your why can help you when picking a career field or branch. For example, perhaps you want to be a pilot. You may at first think Air Force, but each branch of the military has its own fleet of planes. You may have more opportunities to fly serving in a branch other than the Air Force. The largest air fleet is within the Army, and the Navy actually has more aircraft than ships. Another important thing to note is that you have to be an officer or warrant officer to be a pilot. Therefore, if you want to be a pilot, you will need to consider the various officer programs available to achieve that goal. There are also a number of career fields not open to officers. Officers most often serve in management roles. Warrant officers serve in highly technical roles not focused on management aspects. And enlisted members often find themselves doing the hands-on work.

Maybe you are interested in serving in Public Affairs. In the Army you cannot be a Public Affairs officer until after you have served in a different career field within the Army. In the Air Force and Navy, you can begin your career in Public Affairs. Knowing this information could help you determine what branch of the military is right for you.

There are many career variations and rules that apply. Think about what you want to do, then learn more about the career field you are considering.

Do not ever forget that sometimes the only thing you can do is trust your gut. No matter what your why is, and even with all the information at your disposal, sometimes the only way to make the choice is to trust your gut and move forward. Unfortunately, it's not possible to answer every question or doubt before you join, but I have found my gut to be a great source of picking the best option when I have to make a hard decision. Even if you can't quite put into words why a certain branch, career field, or way of service is the right choice for you, it does not mean it is not the right choice. Do your research, trust your gut, and know that if you want to make changes in the future, there is probably a way to make it happen.

FINDING YOUR WHY

Now that you know how important your why is, let's focus on finding your why. People join the military for many different reasons. I asked a few veterans to share why they decided to join the military. Here are a few reasons they gave.

> *To serve, become a medic, and help others. —Laura, US Army*

> *I decided to join because I wanted the chance to serve our country, and I always looked up to those who sacrificed so much that I wanted to be like them. —Lisa, US Air Force*

> *Initially, as a way to pay for college. —Amanda, US Air Force*

I didn't want to go to college right away, and [joining the military] would help pay for my college experience while letting me see the world and serve my country. —Amanda, US Army

My Dad was a Navy chief and my three older brothers were all "lifers" so I thought it was the only thing to do.
—Susan, US Marine Corps

I needed to make a radical change in my life, and joining the Army was there. —Moniek, US Army

I made a promise to a friend that if we didn't make it in Hollywood in a year we would go together as our backup plan. When she saw "a sign" to join that summer, I kept my promise. I had just come back from my brother's Army boot camp graduation and was inspired, so it was perfect timing for her to change the plan. So, a week and one day after my eighteenth birthday we left for basic!
—Brandy, US Air Force

I wanted to escape the life I had before joining. My family struggled financially, and I just really needed some stability. —Isa, US Army

I wanted to get out of my parents' house and get a free education so I could make good money. —Elaine, US Navy

9/11. I wanted to be part of something greater than myself and I wanted to make a contribution for my generation. I wanted to choose a path that would allow me to gain a career and education and be an independent adult. —Jennifer, US Marine Corps

Do any of the reasons these women joined resonate with you? Why do you want to join the military? Service? Education? A change? Tradition? Something else?

Take some time to write out your thoughts about why you want to join the military.

Questions to Consider:

- ★ Why do I want to join the military?
- ★ What do I hope to gain from military service?
- ★ What am I most excited about when I think about military service?
- ★ What am I most concerned about?
- ★ What am I most passionate about?

MAKE THE BEST CHOICE FOR YOU

Joining the military is your choice. No one else can decide if the military is right for you. The military can always say you are not a good fit, but if you meet the standards to join the military, the deciding factor is you.

Other people in your life, such as your family, friends, or significant other, may want to provide input on if you should join, how you serve, or the career field you pick. Whether they think it's a good or bad idea, ultimately you are the one person who will live with the consequences of your decision, including facing the challenges and enjoying the benefits of serving, or the regret or relief of not serving.

Once you have decided if the military is right for you, stand firm with your choice. Often the time between a commitment to the military and actually leaving for training can be months. Do not let a new relationship or the influence of others stop you from joining if you have already decided this is your path. You have gone through the decision-making process, and you know why you are moving forward. Do not let fear of the unknown stop you from taking the jump to serve.

One way you can stand firm in your choice is to find a mentor to help guide you. Another is to listen to the stories of military women. Reach out and connect with another woman who has served in the career field and branch you are working toward. Most will gladly support you in your military service journey.

FINDING YOUR FIT

DECIDING TO JOIN THE MILITARY is the first of many choices you will make in your military service journey. In the beginning you have control over deciding what is right for you. It is in your best interest to do your research as you consider your choices. At this stage, don't give up your freedom to choose to a recruiter. Soon enough the military will be making many of the choices in your life. In the recruiting stage, you still have that control. A recruiter's job is to do what is best for their branch, which may not be what's best for you. So be your own advocate and make the choices that will help you have your best military experience possible.

In this section, you'll learn more about what options are available as you make these choices. Once you understand the options, do more research in the areas that interest you most. Don't worry if you aren't completely sure of your path; trust your gut. Sometimes the best choices can only be made by reflecting on the information you have gathered and taking a leap of faith that you are going in the right direction.

4
MILITARY BRANCHES

THERE ARE SIX BRANCHES OF THE MILITARY that you can serve in: Air Force, Army, Coast Guard, Marine Corps, Navy, and Space Force. Each branch has its own culture and mission, and the military branch you select will shape your whole military experience. There is not a wrong choice when it comes to picking the right branch. Each has its own positive and negative aspects. The right choice for you is whichever branch gives you what you need and want. Start by understanding the differences between each of the branches.

For me, the Air Force felt like the right place to be. I considered the Navy, but my limited swimming skills and discomfort with being out at sea led me to not move past my initial interest. I had looked into serving in the Army and had a negative recruiting experience that left me uninterested. I never considered the Marine Corps. As for the Coast Guard, I did not even know it was an option at the time. People might expect the airplanes are what drew me to the Air Force, but I did not have a love of flying. Instead, when I met the other recruits and was at the recruiting station, I felt comfortable.

Somehow, I knew it was the right place for me. Even with all the research you can do, sometimes you just need to trust your gut to tell you which branch is the right place for you.

You might think "military" means the standard of living will be the same no matter which branch you serve in, but there is a distinct difference in the standard of living (even at war) that each branch is accustomed to. For example, the Air Force is known to have the best living accommodations. The Space Force is its sister service and currently is on par with Air Force standards of living, maybe even better since battles are fought in space from behind a computer instead of overseas. The Army and Marine Corps have the lowest standards of living by most people's standards. And the Navy and Coast Guard land somewhere in between—somewhat nicer at the home installation, but when out at sea, limited space is available aboard the vessel that is your temporary home.

I first learned about the differences between standards of living when I went to training in college with the Army officer program. The Army recruits slept under the stars in sleeping bags on the ground with outhouses for bathrooms. The Air Force recruits slept inside on bunk beds with bathrooms and showers. And although almost everyone at one point or another will find themselves out in the field wishing for a long hot shower, the number of times you are in this type of situation will likely depend on which branch you choose, as well as the occupational specialty.

Another thing to consider when picking a branch is the mission and role of each military branch. As we break down each branch in detail, we will look at its mission and role. Since each branch is so different, you may believe you can only do certain jobs in certain branches, but there is actually a lot of crossover between career fields and positions within the branches. In general, most jobs are offered by each branch, with a slightly different focus related to the mission. That means you have the opportunity to pick the best branch for you based on your personality and aspirations.

When you find a branch that seems like a fit for you, talk with other military women in that branch about their experience.

"Research each branch, the culture, the politics, etc. Talk to those currently in and veterans. Choose a branch that you connect most with their values and mission and feel like you will get the most growth, opportunity, and support. Consider things such as deployment rate, location of duty stations, and duty station rotation." —Elaina Cowan, US Navy

AIR FORCE

The Air Force often gets picked on by the other services and is commonly referred to as the "chair force" because many jobs can be done from behind a desk instead of out in the field. But don't let this stereotype lead you to think the Air Force doesn't spend any time out in the field. Depending on your career field, your office could be at a desk, but it could also take you to places far beyond.

Another common misconception about the Air Force is that everyone is a pilot or works on planes. While the mission is focused on flying, there are many career fields outside of the aviation industry. The Air Force has police, lawyers, intel analysts, and even their own elite teams called pararescues or PJs for short. Besides careers directly related to flying, a number of other jobs are required to maintain a base, including engineering, communications, and logistics.

The US Air Force was established on September 18, 1947. But its roots began in the Army in 1907 when the military created an aeronautical division. They began testing their first airplane in 1908. The early testing led to the US being ready for WWI with the 1st Aero Squadron. World War I and II showed the power of air superiority, and after the war was over, it was decided to create a separate force.

The Air Force's mission is to fly, fight, and win in air, space, and cyberspace. Its rich history and vision guide airmen (gender-neutral term for Air Force members) as they pursue the mission with excellence and integrity to become leaders, innovators, and warriors.

Air Force motto: Aim High ... Fly–Fight–Win

The Air Force consistently has the highest percentage of women compared to the other branches. Some career fields, such as medics and nurses, have a higher percentage of women, while others in the maintenance career field still find women in a small minority. Even though it is the branch with the most women, men still dominate this branch of the military. Expect the possibility that you may be the only woman in your unit.

The Air Force culture is one that respects every member of every rank. Airmen are recognized by their Air Force Specialty Code (job) and rank. The Air Force has members working on technical equipment, and each member is required to meet high standards to ensure the Air Force meets its mission. Young airmen are given a lot of responsibility. In a maintenance shop, the airmen do the hard technical work of fixing and maintaining the aircraft with the support of a shop supervisor. In the civil engineering office I worked in, the young enlisted members were sent out to various job sites to repair or construct new projects. They had a supervisor with them to provide guidance, but they were given the responsibility of completing the task correctly.

The Air Force's mission is focused on getting aircraft in the sky. The base infrastructure must support that mission. Air Force bases are meant to last for long periods of time and because of that, more attention is paid to the amenities. The Air Force builds out a runway and creates its bases with a focus to maintain not only the aircraft, but to also build up those amenities. This means that the living quarters, personal space, and overall amenities, both at home station and in a war zone, are of higher quality than the other branches.

My experience was vastly different when I deployed to Afghanistan with the Army. The Army crammed as many people into a tent as possible. My first day out at Warrior (the outskirts of Bagram Air Base) I was in the bathroom getting ready. The Army woman standing next to me noticed my physical training (PT) shirt said Air Force and said, "I thought they didn't allow your kind out here." I let her know I was deployed with the Army and a look of understanding dawned over her. Going home from the deployment, we were released back to the Air Force and were able to live in the Air

Force quarters on the main base at what was affectionately known as Camp Cupcake. As the name implies, the standard of living on base was a night and day difference from Warrior.

Deployments are typically shorter with the Air Force but vary by career field. The flying community of the Air Force typically deploys as a team and is often on a rotational schedule (typically 90 or 120 days), depending on your base, aircraft, and the operational tempo at the time. While I was in the service, the Mission Support Group (MSG) and the medical career fields had a high deployment tempo as many of the career fields within the MSG (civil engineers, security forces, communications, services, public affairs) and medical field were supporting the mission of the Air Force as well as the Army. Members were regularly tasked to backfill the roles required for the Army. That is how I ended up deploying with the Army. These joint deployment lengths typically range from 120 to 179 days. There are also remote deployments to combat zones and humanitarian responses that last a year.

On the flip side of this, there are career fields in the Air Force that rarely deploy. The developmental engineering and scientist career fields have limited deployment opportunities as their primary mission is at home station. Their role is to create new technologies, which is most often done at home, not overseas. There are still deployment opportunities, but most often deployments are voluntary and deployment orders come down on an individual basis rather than as a group.

ARMY

The largest of the military branches is the Army. Unlike the Air Force's focus on using aircraft to meet various mission objectives, the Army moves primarily by foot or ground vehicle. While military warfare has changed in the last 100 years, the basic principles of temporary structures and a ground combat mission remain the same. The Army needs a large base of support to continually keep troops on the frontlines supplied and moving forward to engage with enemy forces, while also protecting the bases they have established.

The US Army was formed on June 14, 1775. The Second Continental Congress formed the Continental Army as a means for the thirteen unified colonies to fight the forces of Great Britain. George Washington was unanimously elected as commander in chief of the new Army, leading the colonies to victory and independence.

The Army's mission is to deploy, fight, and win our nation's wars by providing ready, prompt, and sustained land dominance by Army forces across the full spectrum of conflict as part of the joint force. The Army is there for the long haul. The Marines are typically the first in and their mission is often quick. The Army follows the Marines after the initial invasion.

All career fields in the Army were opened to women in 2016. Previously, women could not serve in the combat arm branches, which meant women were excluded from the infantry, field artillery, armor, special forces, and some other specialties. Even though the war on terror blurred enemy lines and put women in direct combat, women were excluded from the training men received. The general public saw the story of Jessica Lynch, an Army private captured by Iraqi forces and then rescued by US Special Forces, as a heroic tale, but leaders at basic training in the years following her rescue used her story as an example of failing as a soldier. There is a double standard for women in the Army. You have to run faster and work harder to prove your worth. As women continue to break barriers and glass ceilings in combat arms, it slowly changes the way women are perceived. And while women have an opportunity to excel in the Army and many positive changes are happening for women in the Army, there is a long history of discrimination and unequal footing that women are working to overcome.

Army motto: This We'll Defend

Every member of the Army, no matter the Military Occupational Specialty (MOS) job code, is a soldier first. The goal of the Army is to get the mission done. They are tactical and always moving forward. This focus on moving ahead means that creature comforts (and sometimes necessities) of life are not prioritized. I interviewed women who were in the initial stages of deploying to Iraq, and they talked about living out of their trucks for days as they drove into Iraq after the initial invasion. Laura Colbert was with one of the first units in Iraq after the Army invasion, and they lived in the bombed-out buildings of Saddam's palace. They had to build up temporary structures for bathing, bathrooms, and more. As the deployment continued, they eventually moved from the buildings to tents.

The Army rank structure allows for order and discipline. Many times during my Army deployment training it felt like we were doing repetitive training or doing things backward. As I learned more about the Army, I learned that often the choice to do something is because it is the most efficient or helps build the muscle memory you need. This methodology allows for quick action to be taken once orders are issued and when lives are on the line. The Army is always preparing for the worst-case scenario so it can be ready to respond.

The overall culture of the Army is to train at the most basic level for everyone. A great example is when I went to my combat arms training before I deployed with the Army. I had to clear a weapon (ensure there were no bullets that could accidentally fire) before I went inside a building. During the training we never had ammunition for our weapons, but I had to clear mine every time I went inside. It annoyed me because I knew how to clear a weapon. But at the end of my training, I could clear a weapon without even thinking about it because I had done it so many times. Even though I knew how to do something, I did not have the muscle memory to do it without thinking until after I went to this training. Going through the process of clearing unloaded weapons could have been perceived as a waste of time. But the repetition gave us the muscle memory. The Army's training is based on this approach, teaching at the most basic level and doing the same thing over and over again.

Because the Army is made up of so many people, even if you are the only female member of your particular unit, there are often other women within the larger group. Even as the number of women in combat units is increasing, their numbers remain very low. If you are planning to pursue a combat role, expect to be one of very few and possibly the only woman.

Army deployments are typically twelve to fifteen months.

COAST GUARD

The Coast Guard has a unique mission. It is both a federal law enforcement agency and a military force. This means the Coast Guard has both a peacetime mission and a wartime mission.

Founded August 4, 1790, the US Coast Guard was created when the first Congress authorized construction of ten vessels to enforce federal tariff and trade laws and to prevent smuggling. It has been called the Revenue Marine and the Revenue Cutter Service and has grown in size and responsibility as the nation has grown. It was given its current name of Coast Guard in 1915. Since 2003, the Coast Guard has operated as part of the Department of Homeland Security, serving as the nation's frontline agency for enforcing the nation's laws at sea; protecting the marine environment, coastline, and ports; and saving lives. In times of war, or at the direction of the President, the Coast Guard serves under the Department of the Navy.

The Coast Guard manages six major operational mission programs with the overarching goal to ensure our country's maritime safety, security, and stewardship. The mission sets are Maritime Law Enforcement, Maritime Response, Maritime Prevention, Marine Transportation System Management, Maritime Security Operations, and Defense Operations. If it sounds like the Coast Guard has a lot of roles and opportunities to choose from, you are right.

The Coast Guard also deploys at regular intervals when you are assigned to a ship. I was surprised to learn all the different places the Coast Guard travels for deployments. Whether it's icebreakers in the North or South Pole, traveling in the Pacific or Atlantic for various missions, or deploying with the Navy for an exercise or mission overseas, the Coast Guard's role and scope can extend far outside the borders of the United States.

Coast Guard motto: *Semper Paratus* **(Always Ready)**

If you don't like water or don't want to be on a ship, the Coast Guard is probably not the right branch for you. Even if you pick a career field that doesn't require you to be aboard a ship, that does not guarantee you will not find yourself aboard one anyway. The Coast Guard has its members head to basic training regardless of a school opening. This means that you may end up in limbo after basic training but before you're sent to your trade school. While most often this limbo state ranges from six months to a year, it could last years while you wait for your school to open up. During this time, you are unrated and do the jobs that are open, allowing you to gain firsthand knowledge about various aspects of the USCG. These jobs can range from doing various paperwork tasks to being an entry-level member on a ship.

How often or whether you end up out at sea will likely depend on where you are stationed. Tours out to sea vary depending on the ship you are assigned to. Some missions are response and only last a few hours while most coastal missions are shorter tours ranging from a few days to as long as a few months. There is also the possibility you will deploy with the Navy and those tours are typically six to twelve months. If you are assigned to a ship, your ship will be on a continual rotation of deployments based on the mission. There are different deployment operational tempos, and not all Coast Guard deployments are off the coast of the United States. Sometimes the Coast Guard supports the Navy in the Middle East. The Coast Guard also has deployments not attached to a ship based on the needs of the military.

The Coast Guard has the highest retention rate of all the branches. They do this is by carefully crafting policy, having attentive leadership, and creating programs that work with members and their life situations to progress in rank even if they have to step away from service for a period of time. For example, the Cost Guard offers a temporary separation program for men and women to pursue growth or other opportunities outside the service, while providing a mechanism for their return to active duty. This small branch has a close-knit community.

MARINE CORPS

The Marine Corps is known for being the toughest branch, and this is not a stereotype to take lightly. Passing Marine Corps basic training isn't just something you do; it is a rite of passage. This does not mean the failure rate for basic training is any higher than the Army or Navy. Instead, a willingness to learn goes a long way and the drill sergeants push you hard to help ensure you become the best Marine you can be. If you join the Marine Corps, know it will be tough. Learn all you can about the culture. If being in the branch reputed to be the toughest and most elite is important to you, then the Marine Corps is probably right for you.

Founded on November 10, 1775, the US Marine Corps was created when the Continental Congress ordered that two battalions of Marines be raised for a specific force to work directly with the Navy to perform offensive assaults from both land and water. Marines have fought in all wars of the United States. While the Marine Corps is its own military branch, it falls under the Department of the Navy.

Marine Corps motto: *Semper Fidelis* (Always Faithful)

The Marine Corps has the lowest percentage of women across all branches by a considerable amount. But this stat is trending up. You will stand out as a woman in the Marines, but that doesn't mean you will be limited in advancement. Women have made huge strides in the military and are leading the way in every branch. The Marines pride themselves on being the best. First to the fight is something that is bred into their culture.

Marine deployment is typically six months to a year. Marines also deploy on Marine Expeditionary Units (MEUs) within the Navy. So, if you think you won't be out at sea if you join the Marines, you could find yourself surprised when you end up on a Navy ship.

The Marine Corps is a great branch for people who have something to prove. Marines are tough. If you want to prove you are tough enough, strong enough, and ready to be part of the elite force that is the Marines, this is the right branch for you.

NAVY

A career in the Navy could take you from the depths of the sea aboard submarines all the way to a career in aviation and anywhere in between. In 1993, Congress opened the doors for women to serve on combat ships and aircraft. It took years for the doors for women to fully open because often there were no quarters for women on the ships. In 2010 the Navy lifted the ban on women on submarines and in 2016, doors opened for women to serve as Navy SEALs.

The Navy has four groups, covering the globe. The surface fleet consists of sixteen different classes of vessels for both operational war and maritime response. Roles range from providing insertion or cover for Special Ops forces to battling modern-day pirates on high seas. The submarine fleet is known as the "silent service." Submarines give the Navy the ability to arrive on scene before the enemy is aware they are on their way. The Navy also boasts its own air fleet, consisting of helicopters, fighter jets, surveillance, transport and cargo aircraft, and unmanned aerial vehicles. Along with the war-fighting efforts, there is a team of people back home on what is called shore duty. The shore establishments provide support to the operating forces. Shore duty is often rotated between sea duty and gives sailors a chance to have years of stability at home. Sea duty assigns sailors to a ship, often on a rotation basis for deployments.

The US Navy was founded on October 13, 1775. The Continental Congress decided to arm two sailing vessels with ten carriage guns, as well as swivel guns, and was manned by crews of eighty men. Their mission was to intercept transports carrying munitions and stores to the British Army in America. The founding of the Navy was a pivotal step in the Revolutionary War as colonists first began to fight for rights under the British Empire while some members of Congress resisted the bid for independence. Creating America's own fleet would change the landscape of the war.

The mission of the Navy is to maintain, train, and equip combat-ready Naval Forces capable of winning wars, deterring aggression, and maintaining freedom of the seas.

Navy motto: *Non Sibi Sed Patriae* (Not Self But Country)

Just like the Coast Guard, if you don't like water or the idea of being on a ship, you should probably pick a different branch. Deployments with the Navy are longer than with the Coast Guard and range from six to twelve months. The Navy also has deployments outside of deploying on a ship, based on the needs of the military.

The Navy is second only to the Air Force in having the highest percentage of active service women. *Sisterhood of Mother B* is a website and podcast, created by a group of women Naval Academy cadets, that has resources to learn about the Navy, along with stories of women who have served in the Navy.

SPACE FORCE

The Space Force's role is to monitor and process the established satellite fleet of more than 2,000 active satellites worldwide. The Space Force is not space exploration; National Aeronautics and Space Administration (NASA) is focused on space exploration and the astronaut program. The Space Force is working to create and use new satellites with emerging technologies to maintain space superiority. There is a lot happening within the space arena. Russia and China have their own Space Force. The US Space Force is full of opportunities for those interested in science and technology.

The US Space Force was established on December 20, 2019. The US military's newest branch is still in the process of being formed. The Space Force falls under the administration of the Department of the Air Force. Its conception goes back to 1982 when the US Air Force created Air Force Space Command. Cold-War era space operations focused on missile warning, launch operations, satellite control, space surveillance, and command and control. In 2005, Air Force Space Command gained the responsibility of cyberspace, but this responsibility was transferred to Air Force Air Combat Command in 2018. This allowed Space Command to focus on maintaining space superiority and outpacing our adversaries in the space domain. There are still a lot of unknowns as this new branch forms.

The Space Force is a military service that organizes, trains, and equips space forces in order to protect US and allied interests in space and provide capabilities to the joint force. Space Force responsibilities include developing military space professionals, acquiring military space systems, maturing the military doctrine for space power, and organizing space forces to present to our Combatant Commands.

Space Force motto: *Semper Supra* (Always Above)

The Space Force is a great branch for people who are interested in math and science and excited about being on the edge of a new frontier. It is the newest branch of the military, growing and evolving and changing the world as we know it today.

SELECT THE BEST BRANCH FOR YOU

Now that you know more about the history, culture, and mission of each of the US military branches, here is a list of questions to help you decide which branch is right for you:

- What is your objective for serving in the military? Service, tradition (family of service members), adventure (search for adventure, something to prove), training, college assistance, or something else?
- Are there location/geographic considerations?
- What type of work are you interested in and have a good aptitude or capability to do?
- Which service branch(es) offer the career choice you want? Is there a job or career you can only do in one branch?
- Are you looking for a career in the military or a shorter option?
- Are you looking to enlist or take a path to be an officer?
- What branch personality fits you best?

- What are your expectations about life in the military?
- What do you hope to gain from being part of the branch you are considering joining?

Knowing the answers to these questions can help you narrow down which branch you want to serve in. If the career field you are hoping for is only in the Army, then Army is the obvious choice. But if the career you are interested in is available in all the branches, it will come down to picking the best branch for you. For example, if you are interested in being a medic, all branches have some version of medics. But the specific branch will determine how you serve and what your role will be. It is important to understand how the same careers can be very different depending on the branch you choose.

A good example comes from my friend Ryan. He wanted to be a pilot but he was medically disqualified from being able to fly in the Air Force. He did, however, meet the requirements to fly in the Army. He decided to join the Army because flying was more important to him than which branch he served in.

If the choice is still unclear, reach out for different perspectives from others who have served. Talk to veterans you know about their experiences. Ask them why they picked the branch they served in and then ask them if they would pick the same branch if they signed up to serve again. You might be surprised by the stories you hear. If you don't know any veterans, check out these resources:

Women of the Military podcast shares the stories of women who have served and provides resources from women who are currently joining the military.

The Female Veterans Podcast shares the stories of women who have served in the military.

Army Princess on YouTube has a number of videos to help you decide if the military is right for you, along with resources about picking a branch.

Our Community Salutes supports anyone enlisting in the US Armed Services after high school graduation. Along with providing resources, their communities across the country connect enlisted members from the same area. Each year it holds ceremonies to celebrate graduating seniors as they head off to military service.

Operation Military Kids was created to help inform young adults on the various military careers available to them, as well as to provide guidance on all things military. Their site is made up of resources by military veterans.

If you are on LinkedIn, you can search to find veterans and then reach out to them to ask if they can help mentor you. When you reach out to someone, make sure you take the time to check out their profile and tell them why you are asking for their advice. Is it because they are in the career field and the branch you want to serve in? Are you curious about their career field and want to know more? Whatever the reason, tell them and they will be more likely to respond and help. Don't be frustrated if someone says they can't help or is too busy. Thank them for their time and reach out to the next person. And don't forget you can always reach out to me and I will do my best to help connect you with someone to talk with.

Picking your branch doesn't have to be complicated, but you do need to take some time to reflect on which one feels right to you. Sometimes you will know right away. Other times it will take research and hard work to find your place.

5
ACTIVE DUTY, NATIONAL GUARD, OR RESERVES

THE MOST COMMON WAY TO SERVE in the military is serving full time, commonly known as active duty. But serving in the military doesn't always require a full-time commitment. There are two significant advantages to serving in the National Guard or Reserves. First, you can often pick what unit you are assigned to, giving you freedom in where you live. Second, the time commitment is less than active duty, giving you more flexibility for career options. Being in the Guard or Reserves will not exclude you from the possibility of deploying, but often you will have more control over when you deploy or you may know the rotational operational tempo before you sign up with your unit. No matter what you choose, there are advantages and disadvantages. Learn about the various options of service and pick the one that best fits your needs and wants.

ACTIVE DUTY

Active duty means your full-time job is serving in the military. Once you sign a contract, you will wait to attend basic training. Your wait may be a matter of days, but more often it will take weeks or months before you leave for basic training. Once you leave for basic training, your service commitment begins. Typical active-duty service commitments range from four to six years.

After basic training completion, the next step is either your first base assignment or straight to a technical school for career training. Technical schools teach the ins and outs of the career that will be your full-time job upon arrival at your duty station. Active duty has many benefits, but it is the most restrictive form of serving. You are giving up total control of where you will live once you go on active duty. There will be an opportunity to provide input on what duty station (base/post) you would like to be at, but the needs of the military come first. If you are enlisted, not married, and below a certain rank (typically E-4), most often you will be required to live in the dorms on the base. Married enlisted members do not have to live in the dorms and have the option of living in on-base housing or off-base within the community. Officers are typically allowed to live on or off base.

While active duty has restrictions and requires members to give up a high level of control, the benefits outweigh any cons for many people. Active duty provides a guaranteed paycheck, educational benefits, advanced and specialty training, thirty days of annual paid vacation, tax-free room and board (or allowances), health and dental care, use of commissary and military exchanges, and special home loans and discounts.

NATIONAL GUARD

The Army and Air Force are the only two branches that have National Guard components. There are a total of fifty-four separate National Guard organizations. They consist of each state, the District of Columbia, and the territories of Guam, the Virgin Islands, and Puerto Rico. The National Guard responds to events within their territory and can also be activated to help another state or entity and deploy overseas. National Guard members are controlled by their state's governor, but in 2007 the National Defense Authorization Act gave the President power to activate Guard members without the consent of the state's governor.

Most National Guard members work a civilian job during the week and meet their service commitment by drilling one weekend a month and two weeks a year. There is always a full-time augmentee at each location. Active Guard Reserve (AGR) and Active Reserve Technicians (ART) serve in the Guard full time and receive the same benefits as active-duty members.

When signing up with the National Guard, the unit you sign up with is the unit you are typically expecting to be assigned to. You will still go on active duty for basic training and technical school, but your first (and possibly only) assignment will be with that unit. This allows the flexibility of serving in the military while having control over where you live.

National Guard members can be activated to help during state emergencies, such as national disasters, civil unrest, and pandemics. They can also be activated by the President of the United States to deploy overseas for war. There are laws in place to allow you to complete your yearly training requirement, along with protecting your job if you are activated.

You can switch from one National Guard unit to another and you are not limited to the state where you are currently living. You also can switch from National Guard to Reserves or National Guard to active duty based on life circumstances and the needs of the military. Typical initial service commitments are six years but vary based on job choice, bonuses, and more.

RESERVES

All US military branches have a Reserve component, composed of "citizen soldiers" ready to fight when the nation needs extra members for an overseas response. Reserve members typically serve one weekend per month and two weeks per year, with an initial signing commitment of six years. The main difference between serving in the Reserves versus the National Guard is how members are controlled and activated at higher levels of government. While National Guard members primarily fall under the control of their state's governor, Reservists fall under the President's control. Just like in the National Guard, members will be required to go active duty for basic training and the technical school training required by their job. And the member has control over the unit and location they will be assigned to when returning home. Reservist members can be activated to deploy overseas for combat or humanitarian missions but cannot be activated on US soil due to the Posse Comitatus Act, which was signed into law on June 18, 1878. The act prevents federal military members from being used for civilian law enforcement unless expressly authorized by law.

Similar to National Guard members, members of Reserve forces can transfer to different units around the country as long as there is availability. Reservists are also able to switch to National Guard or active duty based on life circumstances and needs of the military.

The military has many different options when it comes to how you serve, and there is a lot of flexibility to switch from one type of service to another. It is important to understand these differences and know your options so you can make the best choice for you. No matter how you choose to serve, when you sign up, you are doing so knowing that the military has much control over your life and that you could deploy overseas and give your life for your country. Joining through the Guard or Reserves does not take away that reality of military service. It may not be a full-time commitment, but it could require a lot of sacrifice. That is the part of military life that never goes away from serving.

6
OFFICER OR ENLISTED

WHEN DECIDING TO SERVE IN THE MILITARY, people often skip over the step of deciding if they should serve as an officer or enlisted member. Most people join the military by going to a recruiter's office and enlisting. Many recruits do not realize or understand the option of becoming an officer and the differences between these two paths. The purpose of this chapter is not to convince you to choose one over the other. For me, becoming an officer was the best option, and I almost did not find out about it. So I encourage you to learn about both options to see which is right for you. There are advantages and disadvantages no matter which path you choose. Being aware of the differences between officer and enlisted and how those differences might affect your career in the military is an important step in the journey.

The initial qualifications are different to join via the enlisted or officer route. To enlist in the military, you must take the ASVAB and meet a minimum score. Most often you also must have a high school diploma. The Air Force and Marine Corps offer limited waivers for those who have a GED instead of a high school diploma. And if you have a GED, ASVAB score requirements are higher and a higher score will make it easier to get a waiver.

To be an officer in the military, you must have a four-year college degree, pass either the ASVAB or branch specific test with a qualifying score, and be selected to participate in one of the officer programs available.

WHY WOMEN ENLIST

One common reason women enlist is because they feel they are not ready for college. Some women try out college and figure out it's not for them at this point in their life. Enlisting is a great way to get a career started and then use the structure, skills, or tuition assistance to get a degree after completing military service. Others find that even though they are ready to attend college, they do not have the financial means to do so. Even with a military scholarship, it may not be enough to be able to attend college. Instead, they choose to enlist so they can build a foundation and bring in a steady income. They may use their time in the military to build a stable financial situation to help them pay for a degree. If they are focused on completing their degree, they are able to begin taking college courses with tuition assistance while serving on active duty. Others plan to use their education benefits when they leave the military and transition to the next phase of their life.

Women may choose to enlist instead of becoming an officer because not all career fields are open to officers. The career field you may be most interested in may only be available to those who enlist. Many of the hands-on careers are only available to enlisted members while, generally, officers are focused more on management. If you want to learn a skill or trade, enlisting is the best option.

Advantages of Enlisting

One of the main advantages when it comes to enlisting is the hands-on career training you will receive. The military will jump start your career in your chosen field. Depending on how well that role translates to a career outside of military service, it can give you a head start on your post-military life. Many careers in the military give young people more opportunities that would require more experience in a civilian field. Some jobs are not a common occurrence outside the military. The military provides updated training throughout your time in the service at no cost to you.

If you enlist, you join a technical career field. An officer will be assigned to a general area of expertise, but in that one area there will be many different technical job codes. For example, I was a Civil Engineer, but in the Civil Engineer squadron there were a number of different main specialties (e.g., fire fighters, electricians, structures, engineering assistants, environmental engineering, fuels, water treatment, and more). Within these broad groups there were sometimes even more specialties when it came to specific job functions. When I was an O-1 (initial officer rank), I worked alongside enlisted troops to learn more about their jobs. I learned how to solder a pipe, use a front-end loader, build a concrete masonry unit wall, adjust elevation sights, and more. It was fun to be out in the field spending time with the enlisted members in our unit and learning about their jobs, but I never learned the technical aspect. I was a manager. Maybe you want to do the hands-on work. If so, enlisting may be for you.

Another example of a career field is the military band. The military band members are all enlisted (only the conductor is an officer). Charlan, a musician, wanted to join the Air Force Band. She had a master's degree and an extremely high ASVAB score. Her recruiter urged her to consider being an officer, but then she wouldn't have been able to join the band. She won an audition, was selected to be a member of the band, and joined as an enlisted member.

Another key advantage of enlisting is the earned military benefits you gain from service. Once you leave the military, you can use these benefits to help you continue your career. You can go back to college or complete technical training. Military and veteran organizations provide multiple benefits as well. After four to six years of serving in the military, you can use the tools and benefits you earned in the military to continue to improve your life.

Disadvantages of Enlisting

Initial pay for enlisted members is very low. You start at the lowest level rank and pay grade. When you live in the dorms, you get free housing and food from the dining hall, and your base pay only. While it is enough to live on if you are young and single, it is not a lot of money. If you are trying to support your family, you will find the pay allowances for junior enlisted and even mid-grade enlisted members do not go very far. (Military compensation rates are public and available at MilitaryPay.Defense.gov.)

New enlisted members also get less freedom. Even though everyone in the military is subject to rules and restrictions, the newest enlisted members find themselves with the least freedom. The role of the military is to keep everyone in line, so the level of control is extreme to ensure everyone follows the standards. As members increase in rank and complete training, more freedoms are granted.

Enlisted members often are not given all the information necessary to understand the bigger picture. This can also be true for officers, but typically, the higher in rank you are, the more information you can access. By the time the order gets down to the lowest rank, you have to follow orders and do it even if the task does not make any sense to you. Everyone is required to follow orders.

Enlisted members often get stuck doing the job no one else wants to do. Remember the hands-on work I mentioned? Depending on the career field you choose, that may include manual labor.

Allie Braza joined the Navy as a yeoman and expected to do administrative and clerical work at her first assignment. Instead, when she reported to her first ship, it was in dry dock and going through repairs. Her first job was helping with those repairs. She shares:

> "It was all maintenance work. We had a very, very old ship. The paint was lead paint. So, you had to wear tons of PPE (Personal Protective Equipment). And it was a lot of everybody, no matter what your job was, everyone was doing hard labor, like paint chipping. We were cleaning, lots of cleaning the ship down after all the maintenance. We were just doing a complete overhaul of

the ship getting ready to be able to deploy again. So, it was hard, manual labor."

Hana Romer enlisted in the Marine Corps when she first went on active duty. She had never done blue collar work before joining the Marines, but her career field of aviation ordnance required her to be on her feet all day, covered in grease and working with tools. She found herself completely worn out at the end of each day.

MILITARY OFFICERS

Military officers are the leaders of their organizations. Brand new officers outrank all enlisted members. Even with their limited experience they are put in charge and are given responsibility (even those starting very young). Good officers know they need the help of senior enlisted members to help them as they make choices, especially in their early years. Being an officer requires taking on responsibility from the beginning of your service.

Advantages of Becoming an Officer

There are many advantages to becoming an officer, in addition to the guaranteed job awaiting you after completing officer training requirements. If you are able to begin your officer training while at college (ROTC), you may even be able to get college paid for.

One of the biggest advantages is that being an officer gives you the opportunity to be in leadership roles at a young age. The military is known for giving young officers a lot of responsibility early in their careers. Many young people who work in the civilian sector have to work their way up to a leadership position, but many young officers come into the military managing people, logistics, and large budgets from day one.

While the lowest-ranking officer often gets left with responsibilities no one else wants to do, officer jobs involve less grunt work than enlisted jobs. While the youngest officers might end up with the duty of keeping the snack bar full, younger enlisted members might find their duty to be picking up trash or cleaning toilets. It does not mean that officers will not have to get dirty and do things they may not want to do, but often they will have a choice about what tasks to do.

One other major benefit of joining the military as an officer rather than as an enlisted member is your pay is higher. Your pay also rises quickly the first four years you are on active duty. You start at the bottom of the officer pay chart in both rank and time in service, but every two years your pay changes for the time in service and most officers make the rank of O-3 by the time they hit the four-year mark.

Disadvantages of Becoming an Officer

Just like being a leader is an advantage, it can also be a challenge for young people. The training the military provides helps, but it does not change the fact that the military expects a lot from officers, including long hours, lots of training, and time away from family. Being a unit or section leader is rewarding but it is not easy.

Leaders manage the big picture aspect of the work being done. Officers rarely get down in the weeds when it comes to the task at hand. They oversee the projects while relying on enlisted members to gather the information, do the manual labor, etc. This arrangement could be a positive for some people, but I think sometimes when officers first go on active duty, they are excited to get to do something with their hands and then are disappointed to find this is not often the case. The higher an officer's rank, the more paperwork they may find themselves doing instead of being out in the field.

One of the biggest disadvantages of being an officer is how lonely it can be. The rank structure within the military limits who you can spend personal time with, so building close friendships can be challenging. In many units, you may find only a handful of other young officers to hang out with, and this becomes even more true for higher ranking officers. Since the military is male-dominated, the number of women officers is even fewer than the number of male officers. It can be very lonely. Whether you choose the officer or enlisted route, it will be important for you to nurture a strong support system for yourself.

Becoming an Officer

To become an officer, you must be selected for an officer program. There are too many to list here, and rules and requirements are always changing based on what is required by each branch.

One of the most common ways to gain your commission (become an officer) is to go through a program that happens while you attend college. Reserve Officer Training Corps (ROTC) is a common path many branches offer. In ROTC, you take special classes alongside your normal college courses, with various summer training requirements. Other programs only require training in the summer months between classes. Or you can choose the full immersion into military life by attending a military academy.

The specific program you attend and the scholarship you receive will determine how much of your college is paid for. Military academies provide a full scholarship along with a monthly stipend throughout your time at the academy.

If you already have a degree, each branch has its own program and application process for joining the military as an officer.

There are also programs for specialized fields, such as being a doctor, lawyer, dentist, and other professional specialties. Each branch of the military has specific recruitment programs and incentives to recruit members in these specialized fields.

As I mentioned, when you are an officer, you automatically outrank all enlisted members; this gives you the opportunity to lead others from the beginning of your career. And while smart young officers learn how important it is to rely on their senior enlisted members when facing tough challenges, it ultimately ends up being the officer who has the final say. Obviously as you go up in rank, the responsibility of each choice grows and affects more people. But even when you are brand new to the military, some of the choices you make could have an impact on your unit. This is a lot of responsibility for new officers. And it's why it is so crucial to rely on senior members who have been there for years who can help guide you to make the right choices along the way. The relationship between senior enlisted members and young officers is a special one and very different from the one between young enlisted members and senior enlisted members. The senior enlisted leaders who helped lead me to make the right choices and learn how to be an officer often were parent figures and mentors to me. The mentorship that young officers receive in the military from senior enlisted members and senior officers is one of the best parts of being a young officer.

Enlisted to Officer

It is also important to mention that there are opportunities to switch from enlisted to officer through various programs, such as Officer Training School, attending a military academy, or other specialized programs. The expertise of enlisted members who then become officers is invaluable, not only for the military at large, but to help officers who were not enlisted to understand the military better. Some of my greatest mentors were fellow officers who first served as enlisted members before switching to the Officer Corps.

If you want to be an officer, enlisting can be a good option to help you gain experience before you become an officer.

THERE IS NO WRONG CHOICE

Even if I had not learned about becoming an officer and had enlisted, I think I would have had a great career in the military. It certainly would have been a different experience. But I don't think I would have regretted making a different decision.

There is no better option, there is only the best choice for you. The military has so many opportunities within it. The decision on being an officer or enlisted member comes back to knowing your why. Reflect again on why you want to join the military and what you hope to accomplish.

As you discern whether enlisted or officer is the right path for you, consider these questions:

* What is most enticing about enlisting?
* What resonates with you most about being an officer?
* Does the career you want require a degree?

- ★ What kind of work do you want to do on a day-to-day basis? Are you more interested in technical fields and hands-on work or in leading people and organizations? Is the career field you want only available to enlisted members or officers?

- ★ What level of education do you have or want now or in the future?

- ★ Can you picture yourself as an enlisted member or an officer in the branch you're most interested in?

- ★ Which option feels best in your life right now, to enlist or become an officer?

7
CAREER FIELDS

ALL MILITARY OCCUPATIONS AND POSITIONS became open to women in January 2016. In December 2015, Defense Secretary Ash Carter announced that for the first time in US military history, as long as women could meet specific standards:

> "They'll be allowed to drive tanks, fire mortars, and lead infantry soldiers into combat. They'll be able to serve as Army Rangers and Green Berets, Navy SEALs, Marine Corps infantry, Air Force parajumpers, and everything else that was previously open only to men ... And even more importantly, our military will be better able to harness the skills and perspectives that talented women have to offer."

And thus, the military changed and at the same time the military continued as it had always done. The truth was that women had begun to be integrated in combat long before women were formally allowed to serve in these combat units. As the war in Afghanistan continued and the new war started in Iraq in 2003, the military found itself unable to interact with half of the population: Women.

Because of the culture in both Afghanistan and Iraq, American military men could not talk to the local women, and thus they could not get the needed intel this population could provide. American military women began to be embedded into various teams to help with clearing women on base, meeting with women within the local village, and other tasks. While women were still not formally allowed to serve in the combat arms, the military began using the word "attached" to change how women were used in combat.

But women in these specialized teams were just one part of the story of women crossing into combat. As convoys regularly came under attack for various missions, military women continually found themselves in combat situations.

Women often found themselves on the frontlines of war as they worked to help save those in need of their aid when teams came under attack. Convoys rely on the skill and expertise of medics to help save lives during combat operations. Every mission I went on in Afghanistan always had at least one medic in the convoy and two thirds of our medic team were women.

My own experience as a civil engineer in the Air Force led me to believe I would be safe inside a base for my military deployments because I was a woman. But I was sent to Afghanistan with the Army, attached to an Infantry unit, to run convoys throughout the Province of Kapisa in 2010. Before I began asking women about their military experiences, I thought my deployment experience was unique. But the more women I talk to, the more I hear incredible stories about what women have done overseas.

The role of women in the military continues to grow and change. Today's military is working to give women an equal footing.

CONSIDER ALL YOUR OPTIONS

One thing I regret about how I joined the military is the fact that I only ever considered being a civil engineer. My degree was in civil engineering, so it made sense to follow the degree I obtained in college and become a civil engineer in the Air Force. And although I excelled at engineering and loved my job, by the six-year point of my service when I left the military, I had no desire to be a civil engineer in either the civilian or military sector. Engineering was something I could do, but it was not my passion. I never took the opportunity to really think about what I wanted to do with my life.

When I was looking at enlisting, my recruiter only showed me the jobs available that had a bonus. You may face a similar situation. I did not know there were other jobs outside of the ones on the list he showed me. Instead of understanding all the jobs available, I only looked at jobs with bonuses.

My cousin scored high enough to get into nuclear engineering with the Navy. This was not the career field she was planning on but when the recruiter saw her ASVAB score, he pushed her to take the signing bonus and join as a nuclear engineer. In the end, she decided not to join the Navy. Her dad once told me he believed she felt pressure to do a career field she was not interested in and felt she did not have the option to pursue what she was passionate about.

Bonuses are great if your goals and the needs of the military align. But a bonus should not deter you from a certain career field. Your personality and passion should help drive your decision. Remember that in this part of the process of joining the military you are in control. So do not be afraid to walk away from the military or consider a different branch, if the branch you are working with is unwilling to help you get the career field you desire. As long as you qualify and are willing to wait the time required for an opening, you should be able to get the career field you are aiming for. And don't be in a hurry to join if you do not have to be. If you have the flexibility to wait for a high demand career field, choose to wait. Sometimes recruits are in a hurry to get to training, and they take the first job opening with a bonus available instead of waiting for the right career field. Waiting a few months to go to basic training is better than spending years in a career field you do not enjoy. Picking the right field for you will have a huge impact on your career in the military, as well as the career you choose after you leave the military. Do something you enjoy.

WHAT CAN YOU DO IN THE MILITARY?

One question you might ask about career fields in the military is, "What can I do?" A quick answer to that is anything you want. There are so many different career fields within the military. If you dream it, you can probably do it. There is a lot of overlap between careers in civilian life and military life. Plus there are specialized military career fields you likely won't find in civilian life related to fighting a war. The possibilities are endless.

No matter what you choose, there are crossover skills you will gain from your time in the military that can transfer to civilian positions. Some career fields transfer more easily than others. If you already know what you want to do in a civilian career, check to see if that exists in the military or if something closely aligns with it.

For example, if you want to work in the medical field, in almost every capacity the military offers a complementary role. Those who work in the military in the medical career field often do similar tasks they would do in a doctor's office or a hospital based on where they are assigned. A deployment might bring a unique challenge or experience not typically seen within civilian life but the normal at-home station career would easily transfer to the civilian sector when you decide to leave the military.

Even a career field of being an infantry soldier, although it does not correlate directly to civilian life, provides tools that can be applied to a career outside the military. Skills like communication, teamwork, dependability, leadership, flexibility, adaptability, work ethic, and integrity can easily transfer over to the civilian sector in whatever career field you want to pursue. You just have to change the skill you learn from military lingo to civilian terms.

Science, Technology, Engineering, Mathematics (STEM)

All branches have a research and development section where you can be a scientist or an engineer and work on emerging technologies to help keep the military on the leading edge of scientific discovery.

Architecture and Construction

All military branches have operational centers around the world. These operational centers need to be built and maintained. My career field was civil engineering in the Air Force and our mission was to maintain a base to ensure there is a runway ready for airplanes to take off and land on as well as buildings maintained for airplane repair and lodging. This mission was different from that of the Army combat engineers. They were tasked with keeping the Army moving and their focus was on temporary structures, roads, and bridges. The Navy and Marines have their own versions of construction engineers too.

If you are interested in the construction arena, for officers or enlisted, talk with someone in that career field in the branch you are interested in, because they all have a different focus and mission, and you want to align your expectations.

Cyber Security and Information Technology

All branches have some level of cyber security that falls under the US Cyber Command. Cyber security is a growing field and an important asset to the US military. Cyber is also a great career option after leaving military service, with a number of internship programs that can either pay for college or certification with a potential follow-on career. And with the world running on computers, there are information technology jobs in all the branches as well.

Medical Careers

There are a wide variety of careers in the medical career field of the military. The medical team treats and helps military members and their family members in a number of different areas. The main focus areas are behavioral sciences, health administration services, laboratory sciences, optometry, pharmacy, podiatry, and preventive medicine sciences. The training you receive in military medical roles can also easily transfer over to a career outside the military.

Combat Arms

These career fields include but are not limited to infantry, artillery, armor, Army Rangers, Navy SEALs, Green Berets, and Air Force parajumpers. The combat arms are those who participate in direct tactical ground combat. Infantry is the most general description of the combat arms career field. Infantry members fight primarily on foot with small arms in organization with other military units. Certain military aircraft and ships fall under combat roles, determined by their mission and capabilities. Combat arms is an area that has opened up to women over time. Today, women continue to join these career fields and succeed.

Intelligence and Readiness

Intelligence is crucial for our national defense. Critical information about enemy forces, potential battle areas, and combat operations mission support may be gathered from sources like aerial imagery, maps, geospatial intelligence, or interrogation. Intel units may conduct counter-intelligence operations and individuals might specialize in a specific area, such as imagery, signals, or human intelligence. Intel work sounds exciting, but it can also involve a lot of looking at data in a room with no windows, analyzing the data, and using that data to help the military have a tactical and operational advantage. All branches rely on the timely gathering, accurate interpretation, and effective distribution of intelligence.

Finance and Contracting

The finance career field involves planning and providing services for finance related issues within the military. This can be as basic as working out pay issues for military members or as complicated as sourcing money from Congress for various projects and the general maintenance of a military installation. Contracting officers set up contract bids and awards with government suppliers to procure supplies, services, and construction. They lead the contract administration, handle purchase transactions, and ensure proper paperwork is in place to comply with federal regulations.

HR Managers/Military Personnel Office

Like a human resources manager in the civilian sector, military personnel managers ensure members are on track for promotion, training programs, and job specialties. They also track and maintain their health records.

Law, Public Safety, Corrections, and Security

Members in this career field help enforce military laws and regulations. They also provide emergency response to incidents on base or to disasters. This includes firefighters, military police, and other law enforcement and security specialists. Lawyers, judges, and the whole of the military justice system also fall under this category.

Transportation, Distribution, and Logistics

Covering land, air, and sea, transportation responsibilities include transporting supplies, cargo, and personnel. The military has a huge logistics infrastructure to track everything it manages around the world.

Media and Public Affairs

Members in media and public affairs prepare and present information about military activities to the military and the public. They have a number of different responsibilities ranging from taking photos, writing articles, creating videos, and managing media contact, just to name a few. Training for both officers and enlisted for all branches is at the Defense Information School.

Lastly, I want to talk about the danger related to various career fields. No matter what career field you choose when joining the military there is a possibility that you might be injured or asked to give your life for your country. But some career fields, by the nature of the work they do, have more inherent risk based on factors like frequency of deployment and the missions related to each career field. Some of the most dangerous career fields in the military are rifleman, pilot, combat medic/corpsman, explosive ordnance disposal expert, cavalry scout/reconnaissance, pararescue, special forces, infantry, and combat engineer.

Now that you have a general understanding of what career fields are available, it is your turn to do the work and figure out what you want to do. Take time to write down what you are passionate about, things you don't like, what you are hoping for. Use these questions to help you decide.

Questions to Consider:

- If you could do any job in the world, what would you do?
- Which jobs have bonuses (if that is important to you)?
- Does this job have an additional service commitment?
- Do you have to be an officer or enlisted member to do this career field? If both officers and enlisted members can choose this field, how do the roles differ?
- How long is the technical school for the job you are considering?
- To what duty stations can this job be assigned?
- What is the deployment/op tempo of the job you are considering?
- How can you use this job when you leave the military? Is that something you would be interested in doing as a civilian?

There are many tools you can use to learn more about your strengths and skills. CareerGirls.org offers a personality-based quiz to see which careers might be a good match for you.

8
EVALUATIONS

THE MILITARY REQUIRES ITS MEMBERS to meet basic standards to serve in the military. In this chapter, we will dive into two basic qualification factors required for all military members: a standard entry test and a medical physical. It is important to be prepared for both of these events. Preparing for the military tests can help ensure you get a high score. And knowing what medical disqualifications could affect you can help you determine if you can serve in the military and begin the waiver process before you arrive at the Military Entrance Processing Station (MEPS).

MILITARY TESTS

Most members who join the military are required to take the Armed Services Vocational Aptitude Battery, commonly known as ASVAB. There are three different versions of the ASVAB test. The CAT-ASVAB is a computer-based test given at MEPS. The MET-site ASVAB is a paper-based test given at satellite locations. And the Student ASVAB is given by schools for career exploration. The tests have slight differences in length and time cap, but in general each version takes around two and a half hours and has nine sub-tests. Practice tests can help you prepare for the test. No outside help is allowed, including no calculators to help with the math section.

While you can take the CAT-ASVAB at MEPS, if you have the option to take it on a different day, I recommend trying to go that route. It might seem more convenient to take the ASVAB and complete the MEPS medical physical on the same day. But it can also add more stress to an already stressful process. If you are enlisting and pass your medical physical, job assignments are regularly decided and finalized at MEPS. Sometimes when members take the ASVAB and go to MEPS, they are required to pick a career field the same day they learn what career fields are available to them. This often leads to members picking a career field based on the bonuses offered and not on the desires of the member.

Your ASVAB test helps determine your Armed Forces Qualification Test score, which is reported as a percentile between 1 to 99. So, if you get a score of 90, it means that you scored as well as or better than 90 percent of a nationally-representative sample of 18- to 23-year-olds. If you score a 65, then 35 percent of the test group scored as well or better than you. The score is used to help you see if you are eligible to serve in the branch you are looking into. It also allows you to see what jobs you are eligible for. Some schools use the score to help young people determine what career is best suited for them. So even if you do not join the military, you can use your score to help you determine a career path for the future.

Some military branches accept the ASVAB for officer applicants while others require different testing, such as the AFOQT (Air Force) and ASTB-E (Navy). Your recruiter can provide you more information about testing requirements if you are considering becoming an officer.

It is really difficult to study for the ASVAB test. Luckily, there are a number of practice tests available to help you get familiar with the types of questions and give you a general idea what to expect. You will not be able to study for each category but knowing what to expect is a great way to be ready for the test. The most important things you need to do to prepare are to get a good night's rest, eat a big breakfast, and relax. Then just do your best. The majority of people score high enough to serve in the military; the higher your score is the more jobs you will have available to pick from.

MEDICAL PHYSICALS

Everyone who serves in the military must go through a medical physical. All enlisted members go to Military Entrance Processing Station (MEPS) to be processed for a medical physical. Most officer candidates go through the Department of Defense Medical Examination Review Board (DoDMERB) in the process of applying to an academy or ROTC and will require a final physical set up by the military. My final physical was at MEPS. The military medical physicals are an in-depth physical that goes through your medical history in written form along with a thorough medical physical. The DoDMERB physical is a multiple step process that takes weeks or months to be cleared for service. A MEPS physical with no issues will clear you the same day. However, should any issues come up, you could find yourself needing to request a waiver or see the end of your military career before it has even begun.

The best way to prepare for the medical evaluation is to ensure your recruiter is aware of any medical issues that may cause you to not pass the military physical. A good recruiter will go over common disqualifications when you meet with them and should be able to prepare you for what to expect at the screening. Be sure to be open and honest with your recruiter about anything you are concerned about. There was a time when your past medical history was difficult for the military to obtain, but in the age of digital data it is much easier for the military to check your records and ensure your records match. It is better to talk to your recruiter in the beginning of your military journey instead of finding out at MEPS that you are disqualified for military service. Some waivers can even be completed before arriving at MEPS.

But MEPS is more than a medical physical. They also check your criminal record, screen for drugs and alcohol, and review your record for past military service other than honorable or if any errors are found in your medical record.

DoDMERB Physical

This physical will vary by your location. When I went through the DoDMERB physical, everything was set up by the ROTC leadership. If you are applying to a military academy, there are instructions on their website for how to schedule your own DoDMERB physical. The clinic I went to was in my hometown. They were certified to do the physical. It was a quick appointment and reminded me of a standard sports physical, just more in depth and with a lot more questions to answer.

Going to MEPS

MEPS, unlike the DoDMERB, is a more time-consuming process. When it is time to head off to MEPS, you will need to bring certain things with you. Everyone is required to bring their social security card, birth certificate, and driver's license (if applicable). If you wear glasses or contacts, bring them with you along with your prescription and lens case. Make sure you dress for the event. It doesn't have to be formal, but tank tops and ripped jeans are not going to fly. You should also plan to be there all day. Packing a backpack full of snacks and water is a great idea. You may have to wait for hours after completing the physical to meet with the detailer (the person who helps set you up with your career field). You don't want hunger or thirst to hurry your career field decision. Make sure you are prepared so you can be in the best state of mind and have the energy to fight for the career field you want.

Most often potential military members arrive at a designated hotel in the afternoon the day before their physical. There is an option to arrive at the MEPS facility in the morning, but if you are late for any reason, you could miss the start of the physical and not be allowed to be screened. Once you arrive at the designated hotel, you check in with the military representative and get assigned to your room. It is not uncommon for military recruiters to bring a group of military candidates who either are shipping out for training or being processed for medical physicals. You may be required to share a room with another recruit or may get lucky and have your own room. The military representative will give you instructions on where to meet in the morning and what time to arrive.

On the day of your physical, you board a bus to the MEPS facility and then start the process of IDs and medical information by filling out all the required forms. Once you sign all the paperwork, everyone will be sent to various stations for medical screening.

The general physical includes blood pressure, heart rate check (must be below 100 beats per minute), hearing screen, visual screening (depth perception, color blindness, overall vision), reading test (AF), and a blood draw. Each person is also required to take a breathalyzer test. After these initial tests are done, the group is divided into men and women for a full body physical.

When you change out of your clothes for the screening to a bra and underwear, you will also be required to provide a urinary sample to check for drugs and ensure you are not pregnant. Next, they will check your height and weight and then test your range of motion. There are up to twenty-five various range of motion exercises used by the military, the most daunting being the duck walk. I recommend learning what the duck walk is and practicing it at home before arriving at MEPS. It is stressful to follow the quick instructions of the military representative and move in an awkward motion.

Lastly, you will have a full body physical with the doctor. Each member will go into a small exam room one at a time and be asked to take off their bra and underwear and put on a hospital gown. The doctor and a woman supervisor will be in the room as the examination takes place. The full body physical consists of checking your heart rate, examining your spine, ears, nose, and throat, etc. It also includes a breast exam and the doctor checking your butt, vagina, and urethra. I was not made aware of this section of the physical and found it uncomfortable. It was quick but caught me off guard and is one thing I wished I had known about before going to MEPS.

With this final step you will either be cleared for medical service, disqualified from service, or put in a hold status for waiver requirements.

Final Process to Leave MEPS

If you are an officer candidate, most likely you will receive the necessary paperwork and be free to leave. But if you are enlisting and have cleared the medical physical, most often you will be sworn into the service. Sometimes this will also finalize your career field. Other times this will happen later in the process with your recruiter. No matter when you sign for your career field, it is important to stand up for yourself and fight for the career field you want. If you have already agreed to a career field before arriving, make sure the detailer honors that request. Do not be afraid to walk away or feel that you have to sign something if it is not what you want. Remember, the recruiters work for the military and they want to fill the roles they need filled.

Some members will go back to MEPS the day before they head off to basic training for a final weigh-in and in-processing.

BASIC TRAINING

YOUR MILITARY CAREER BEGINS with basic training. Each branch has its own variation of initial training for enlisted members and officers. This intense training will break you down and build you up into what the military needs you to be.

Some people believe it is impossible to fail basic training, but the truth is about 15 percent of recruits who join the military every year fail basic training, and their plans to serve in the military end. Most often when people fail it is because they remove themselves from training and quit. It can be because they are not mentally prepared, find the training too difficult, or get injured or can't complete the physical training requirements. Members who are injured or do not pass required training requirements can be recycled and start the training over again at Day 1.

Basic training is often broken into three parts. The first part is the shakedown and it begins right when you get off the bus and begin your in-processing. The shakedown consists of a long day (or days) of standing at attention or parade rest as you move from station to station to get every aspect of your checklist complete. And while it may seem like the military is using the in-processing procedure as a way for the drill sergeants to find you and "welcome" you to the military, it actually is an efficient way to get everyone who arrives at basic training through the process as quickly as possible.

The second phase of training will build on what you have learned so far but will also add in an aspect of combat training and readiness. You will begin to learn more about the branch you are serving in along with being trained in skills required for your career.

The last phase of training takes everything you have learned from your training and begins to put it into action through various exercises or training opportunities. You will use the skills you have learned to work together as a team and succeed. You will also be required to pass a final physical fitness test to graduate.

9
MENTAL PREPARATION

BASIC TRAINING IS MEANT TO BE CHALLENGING. Even those who excel at basic training often struggle in one aspect or another. Being afraid of the challenge is completely normal. You might struggle to survive each day and just graduate, or you may excel. Everyone starts out on the same footing.

This chapter is focused on the mental aspect of basic training because it was the area I struggled with the most. Looking back, I had the potential to excel at basic training. Physical fitness was easy for me. But my lack of self-confidence inhibited every other aspect of my training.

The program you are going through will determine the specifics of what you need to bring with you and how you will get there. Your recruiter will have all the information, and there are too many different trainings or variations to cover in one book. These two chapters are a guidepost for all trainings, pulling out the common themes to prepare you to not only get through training but to thrive.

The one thing you need to know is: you can do it. Having confidence in yourself will make the training so much easier. It is okay to be nervous or afraid and to worry about the unexpected. That is normal and a level of anxiety is good because it is part of the experience. But not believing in yourself or doubting your ability to complete training will add an unneeded level of stress. So, start telling yourself now, "I can do this."

THE FIRST DAY

I actually missed my first day (Day 0) of basic training. The flight out of my hometown was delayed, which caused me to miss my connecting flight. When we arrived at the base to in-process, only four personnelists were there. There was no yelling or standing at attention. Everyone was tired and ready for bed. Two women weighed me in and checked my waist measurement to ensure I met the military standards. They looked over my paperwork, then sent me to wait in the lobby. Eventually I went to my barracks and fell asleep.

The next thing I heard was "Get out of bed! Hurry up, we are already late!" Pots and pans banging, music blaring. I woke up confused for a moment and then quickly remembered where I was. I jumped out of bed, grabbed my shoes, fumbling with the shoelaces as I tried to move as fast as I could. "Go here, go there." Confusion filled the hallways as we ran, not sure where to go. Eventually we all made it outside, lined up and ready to head to the parade field for morning formation. All day, I felt completely behind my peers; they had already unpacked most of their gear, but I was just starting. I had studied. I had trained. But in the rush and confusion of first-day instructions, I felt overwhelmed. I felt like I was already failing. Instead of realizing everyone was struggling to adjust, I focused on my failures, and I took this feeling and let it weigh me down.

The first few days are supposed to overwhelm you; that feeling of failure, confusion, and exhaustion is part of the training process. You just have to get through it and know it is part of the game.

THE GAME OF BASIC TRAINING

Everyone who joins the military goes through some type of intense training designed to break them down so they can be part of a team. There is a game and strategy behind this tactic that the military uses for each new recruit. One of the best ways to prepare for your training is to understand this approach. When the drill sergeants yell, add intense pressure, or provide negative feedback, it can feel personal, but it is a tool the military intentionally uses to stress you out. Just keep your head down, respond correctly, and keep pushing forward. Don't think you can do anything perfectly at basic training. When you are criticized, take it in stride and know it is part of the process.

Fighting the game wastes energy. Do not argue if you have been wronged. Just say, "Yes, Sir/Ma'am" and work to correct what they have pointed out. Remind yourself it is part of the challenge.

Having the wrong type of attitude can create extra pressure. Always be humble and teachable. There is a huge difference between being prepared and being cocky. Be willing to listen and learn. As much as you prepare for training, you cannot know everything. Be prepared but also ready to learn and willing to work with your teammates. Be humble, yet confident.

Remember in every challenge that you are training for war, the ultimate mission of the military. Learning to rely on your wingman/battle buddy (the person to your left or right) is part of that training. When you serve in the military, the mission comes first, and you may have to work with someone you do not like or do not know. You also have to learn how to follow orders; in dangerous situations there isn't time to question leadership. You begin to learn these aspects of the military in basic training.

TAKE RISKS

Don't be afraid to take risks while at basic training. There are boundaries in place to protect you if you fail. Not hitting a standard or falling short of the mission may feel like a mark against you, but training is about learning. Failing during training allows for teachable moments. When you take risks, you often find that you can do more than you ever expected.

One of those risks might be stepping up for various roles and responsibilities. There is mixed advice on volunteering. Sometimes different opportunities can give you extra responsibility that may make the experience harder. But it can also give you an opportunity to grow. Looking back at my training, I was so afraid to volunteer and do it wrong that I rarely volunteered. But then I was chosen to be a road guard leader (someone who stops traffic so the flight can run without stopping). I found it was easier than running with the group in formation and having to be in step and running at a slow pace. It did require extra responsibility, but also took me out of the formation and away from the drill sergeants who were focused on the formation and not the road guards. The rest of the training, I quickly volunteered to be a road guard for each long run. Look for opportunities to volunteer. You will be at the training until it is over, so why not make the most of it?

COME PREPARED KNOWING THE BASICS

Besides mental toughness and mindset, do your research before you arrive to training. Becoming part of a military branch requires you to learn a whole new language of terms and even cultural expectations. The more you can learn early on will make this shift easier. Learning things like the phonetic alphabet (Alpha, Bravo, Charlie), your branch's rank structure, branch song, branch creed, and any other military terms or culture will make basic training easier and will give you confidence.

If your recruiter gives you a list of things to learn before arriving, memorize it. Don't think you will have time at training to learn it. Not taking the time before training to learn it will just add an additional stressor to your life. Even with preparation before you arrive, there will always be something new to learn.

STRENGTHEN THE TEAM

Basic training teaches you to rely on others. No one gets through basic training on their own; it is all about team and helping the weakest link. If you are good at a certain aspect of basic training, you can use your skill to help others on your team. Search for what each team member's strengths are and then use them. Our differences make the team stronger. If you see someone struggling, work to connect with them and help them where they struggle. Making your unit stronger helps everyone on the team succeed.

The Armed Forces often glorify loud and outgoing leaders, but everyone has value in the military, especially at basic training. One of the reasons I worried about completing training was because my personality is more reserved. I thought I needed to change my personality to succeed. I tried to be tough and loud and could not do it. It was not me. And when I could not be the person I thought I should be, it lowered my self-confidence. Instead of worrying about what I could not do, I should have focused on my strengths, like my attention to detail, natural athletic ability, and awareness of the struggles of others. Had I focused more on my strengths and not on my weaknesses, it would have made my experience easier. Be the best version of you; you don't have to be someone different to succeed in the military.

JUMP

You will be forced to do things you don't think you can do. "When you come to a great chasm in life ... jump, it isn't that far." Those were the words my commander told me when I left for my deployment to Afghanistan. It is a modified version of the quote by Joseph Campbell, "As you go the way of life, you will see a great chasm. Jump. It is not as wide as you think." Those words became my lifeline for the deployment. Regularly, I had to do something far outside my comfort zone and was forced to jump. And I did so willingly because I hoped it wouldn't be as big of a jump as I expected. And it is true. Each time I jumped it was never as much of a stretch as I imagined it would be.

You might be like I was, leaving home for the first time for this crazy adventure. Don't be afraid to jump. Allow the military to push you and get the most out of this experience.

Your mind is trained to keep you safe. Going away from pain and challenge is a natural instinct. When you face a fight or flight circumstance, push past that flight instinct and find the fight in you to get through the training.

One of my favorite examples of when I learned to jump was at my deployment roll-over training. In this training, they stick you inside of a Humvee wearing all your battle rattle (helmet, combat vest, etc.). The Humvee is attached to a mechanism that rolls it over and over until the vehicle stops upside down. I was freaked out about being stuck upside down. I imagined the blood rushing to my head and struggling to push my body weight (plus combat gear) up. How could I do all this upside down with one hand while unbuckling my seat belt with the other? It felt impossible. I had set myself up for failure by telling myself I could not do it, but I had to climb in the Humvee anyway. I hoped for a second there was a way someone else on the team could help unbuckle me.

I yelled, "I can't get out of my seat. I tried and I couldn't do it." "Well, I can't get to you. There isn't enough room to move around. You have to unbuckle yourself; it is the only way."

Those were the words I needed to hear. The faint hope that someone would help me out of the situation was gone. As I was dangling upside down, I realized I had to be the one who pushed my body up and unbuckled my seat belt and to my surprise, I did it! What I learned from each experience was that I was so much more capable than I ever expected. Don't limit yourself by your own perception of what you can or can't do. Let the military push you past that so you can grow into who they need you to be.

GO IN WITH CONFIDENCE

I was so terrified of getting sent home from training that I did not go in with confidence. The truth was I knew what I needed to do. I was physically prepared and had worked to memorize all the required material before arriving. But my fear of failure overwhelmed me. I didn't believe in myself. Don't let fear of failure overwhelm you. I already know you are dedicated because you are reading this book. You are capable. You can do anything you set your mind to. And you can do this.

Words of affirmation and believing in yourself can help you get through this. Use these words or something that fits your personality to help you prepare your mind for what is to come. A daily affirmation of, "I can do this. I am going to excel in basic training" can change your mindset. Instead of going into basic training with fear and dread, you can be confident and ready.

It seems silly that a daily affirmation can change your experience. But to achieve anything you have to believe in yourself first. My peers at basic training who were ranked at the top came into training with confidence and a goal to be the best. They were not worried about graduating and were instead focused on how they could excel. Those just hoping to graduate ended up in the bottom, like me. Mindset is so important. Everyone can get through basic training, so don't limit yourself to just surviving; go into it prepared to thrive. And it doesn't matter if you excel at basic training or just barely get through it. Everyone who graduates gets a fresh start at the next training. Take the lessons you learn from basic training and do your best. No one can ask for more than that.

I struggled through my basic training and did my best to stay hidden. I often felt I offered little value to the team. But in the last week of training we were sent to the field, and in the command tent I created an organization system. Other members of my team could not believe how quickly I responded to the challenge and how valuable my skill of observation could be. During the last week of training I found my confidence and began volunteering and speaking up.

Don't be like me; know your value from day one, even if you do not know exactly what your hidden talent is. Know that you have something to offer. Take risks, learn from your failures, and use your time at basic training to grow into a stronger version of yourself.

10
PHYSICAL FITNESS AT BASIC TRAINING

BUILDING A FOUNDATION OF PHYSICAL FITNESS is important to help your body be more prepared to face the strenuous challenge required for training. It will also help prevent the likelihood of being injured. Getting shin splints or muscle strains is easy at basic training because of the physical requirements. Don't wait until you are at basic training to get in shape for your training. Start preparing now so you can be ready.

Talk to your personal physician before making health and fitness changes. The resources and recommendations in this book cannot replace medical expertise. It is always important to listen to your body and to ensure that you are physically ready for the program you choose to follow.

HEIGHT AND WEIGHT STANDARDS

One challenge related to fitness before arriving at training is meeting the weight standard.

If you are trying to lose or gain weight, the most important focus should be on what you are eating. Changing your diet can have a huge impact on your ability to fall between the maximum and minimum weight standards. Consider meeting with a nutritionist to help you adjust your diet to meet your goals. It is not as simple as eating more or less food. The types of food that go into your body are key to your success to get to your required weight in a healthy way.

You should always work with your own physician to determine what's best for your specific health needs. The Mayo Clinic offers this advice for gaining weight: Aim to eat five to six small meals every day. Choose nutrient-rich foods such as whole grains, fruit, veggies, lean meat, peanut butter, cheese, dried fruits, or avocados. Try smoothies or shakes instead of juice or soda. Don't go overboard on sweets. If you are trying to lose weight, focus on eating protein, healthy fats, and veggies. Cut out soda, coffee with cream or sugar, and juice. Stay away from sugars.

Make sure you are exercising. Working out can help you both gain muscle and lose fat. The best type of workouts for you will depend on if you are trying to lose weight or gain weight. Cardio can help increase metabolism and help you lose weight. Strength training is great for building up muscle and can help prepare you to gain weight. Finding a healthy balance of both is most desired, but as you approach your official weigh-in, it may be important to focus more on one area to meet height and weight standards. Getting past the initial weigh-in is key to starting your military career.

PHYSICAL FITNESS

The more you do to increase your physical fitness before you leave for training the more it will help you as you go through training. Working out will build a foundation you can build on. Even those who are physically fit find the rigor and training of basic training challenging.

At a minimum, prepare yourself by being able to pass your branch's physical fitness test. Each military branch has its own set of standards when it comes to testing physical fitness. You can look up your test requirements online or ask your recruiter what the highest and lowest standards are to help you prepare to not only pass each test but exceed the minimum expectations. Also, concentrate on building your endurance through running and walking. You will march everywhere you go. Added to the other physical requirements of training, this can have a big impact on you. Many people struggle through training with shin splints or other various ailments that make even walking, yet alone running, challenging. Also, be prepared to complete at least one long run or ruck march during training. Ruck marches can be challenging for women because of the amount of weight required. If you know a ruck march is required, add that to your workout regimen before leaving for training. The best way to pack a ruck for women is to ensure weight is distributed at the waist and not the shoulders.

Always remember, to get through basic training you do not need to exceed the expectations for physical training, you just need to meet them. Bodies are different. Find what works for you, focus on that, and build to get stronger. Building up fitness takes time. Do not be discouraged when you start working on your physical fitness if you do not see results right away. Just work to run or walk a little bit farther, do one more push-up, and over time you will see results.

Pay attention to your body as you begin to work out. While at home preparing for training, rest when you notice pain. Work out other muscle groups, and if the pain does not go away, make an appointment with a doctor. While at training, go to Sick Hall and report pain that makes it difficult to perform basic tasks. Being sore is a part of training, but pain that makes it difficult to walk or run could be something serious. Reporting injuries and sickness at the beginning can help stop them from growing into something serious that could end your military career before it even gets started. There is a difference between pushing through something because it is challenging and pushing through the pain of an injury.

Push-Ups

If you are like me and had never done push-ups on your toes

before you started thinking about joining the military, start by perfecting your push-up skills on your knees. Building the strength in your arms and chest is challenging and takes time.

Focus on form from day one. Military regulations for correct push-ups require your arms to make a ninety-degree angle. It is generally easier to do standard push-ups where your hands are a little wider than shoulder distance apart and your elbows go out to the side. One thing I didn't learn for years was that having your feet together makes push-ups harder. You can have your feet about shoulder distance apart (up to 12 inches). Keep your body straight from shoulders to ankles. Lower your whole body until your upper arms are at least parallel to the ground. Extend your arms completely when raising back up to the starting position. Use a mirror to check your form. The more push-ups you do, the easier it will be to hit that correct mark.

When I started training to do push-ups, my goal was to be able to do twenty of them on my toes. I started by doing twenty push-ups on my knees each night before I went to bed. At the end of the week, I did as many push-ups as I could on my toes. The first week I did two push-ups. The next week I continued to do push-ups each night but now instead of twenty push-ups on my knees, I did the first one on my toes and then did the last nineteen on my knees. I continued the same schedule each week. Some weeks I would add one to five push-ups on my toes and some weeks I would repeat the same interval I had done the week before. At the end of three months of doing this almost every night, I could easily knock out more than the twenty push-ups in a minute that were required to meet the minimum push-up standard for the Air Force. The month before I left for training, I did push-ups every other night for a minute.

Other people like to build up strength by building it into their daily routine. Having three to five set times during the day when you do ten to twenty push-ups can help build up strength. Use the same principle as in the last example. Start on your knees with ten push-ups and slowly build your strength. You will see results over time.

Sit-Ups

In general, the military has a standard for the correct way to do a sit-up. You must have your arms across your chest and your elbows must touch your knees each time you go up. You can use a bar or have a person hold your feet so they do not move. Over the past few years, branches have begun to change the sit-up requirement to a crunch, reverse crunch, or even a plank instead of a full sit-up. Find out what your branch standard is and use that to help you prepare for training. Whatever requirement your branch has is what you should use to prepare for the test.

It took me a long time to gain the core strength to get the maximum number of points for the sit-up portion of the test. I trained for them the same way I prepared for the push-ups. Every other night before I went to bed, I did sit-ups for a minute. Over time I could do more sit-ups each night.

Running

Each branch requires a timed distance run ranging from one and a half to three miles. Building up your endurance to pass your test is important, but another benefit of slowly building your endurance with running is that it can also help prevent injury. A common injury at basic training is shin splints. The best way to treat shin splints is not running and icing your shins. While at training, not running would likely keep someone from graduating, so many recruits push through the pain and risk the possibility of further injuries, such as stress fractures.

Shin splints often occur from overuse. Building up endurance before you arrive is the best way to prevent shin splints.

When you begin training to run, start with what's best for your current fitness level. If running a mile seems impossible, start by building your way up to walking a mile. Or start by running as far as you can, then run for thirty seconds and walk for sixty seconds, until you finish your distance for the day. Once you can run a mile without stopping, work to increase your distance to whatever distance your fitness test requires.

To build strength and improve your overall run time, you should have the goal to run one mile farther than your required test. You can also build in weekly sprint training exercises to help build up your speed. My favorite sprint exercise is sprinting for thirty seconds with a sixty-second break of either walking or slow jogging. Try this for half a mile, then build up to the distance you are training for. If you can do the thirty seconds/sixty seconds without a problem, increase your sprint distance to forty-five seconds with a forty-five-second break.

Pull-Ups

The only branch that requires pull-ups in its physical fitness test is the Marine Corps, but doing them regularly will help you prepare for any basic training. Pull-ups build up muscles in your back and can help make you stronger overall. I never needed to do a pull-up on active duty, but it is something I have incorporated into my fitness routine. I use a pull-up bar that fits in the door frame and a chair to do assisted pull-ups at home.

If you want to build your strength for pull-ups, start with holding the bar for as long as you can. Once you can hold yourself on the pull-up bar for a minute, start working on doing assisted pull-ups. If you are at a gym, find a machine to make assisted pull-ups possible. If at home, use a chair or you can purchase pull-up assist exercise equipment. Eventually, start working to do one pull-up without assistance and slowly build over time. Women are required to do one pull-up to pass the Marine Corps fitness test.

Evolving Military Fitness Standards

Always check current requirements for your branch's physical fitness standards. In 2022, the Army officially changed its physical fitness test to be more combat focused and added exercises such as leg tucks, max dead lift, and standing power throw. Also in 2022, the Air Force began giving its members alternate exercises, such as a 20-meter shuttle run instead of the traditional 1.5 mile run, hand release push-ups instead of traditional push-ups, and an option for cross leg reverse crunches or planks for core strength. In 2020, the Marine Corps started allowing planks in lieu of timed crunches, in anticipation of the official change in 2023.

GENERAL HEALTH

Another added benefit of being physically fit before you arrive will be that it gives your immune system a boost and allows you to stay healthy throughout your training. You do not want to get sick while at training. It can make the training more challenging and make it difficult to meet standards. If you get a sickness causing you to miss a certain number of days of training, you may be required to start the training all over again. Remember to wash your hands, drink lots of water, and get sleep each night.

Hydration

Speaking of water, you will drink a lot of it, especially if your training is in the summer months. Ensure you are getting enough electrolytes too. At meals you often are given a choice of a drink that has electrolytes in it. Even if it is not something you prefer, you need those electrolytes with all the exercise you are doing to help fuel your body. Water is great but it is not enough. Drinking the juice or sports drinks provided is critical to help you stay strong during your training. Dehydration can be a real problem at basic training.

Eye Care

If you do not have perfect vision, plan to wear military-issued glasses at training. Gone are the days when the military issued what were commonly referred to as BCG (birth control glasses). Now the military has more modern frames. At basic training you will be issued a pair of glasses within the first few weeks of training.

I usually wore contact lenses in my civilian life, but I brought my glasses to basic training, as required. I wish I had brought my contacts as well. Some recruits brought theirs and sometimes were able to wear them. Do not plan on wearing contacts the whole time, but you can bring them in case you have the option to wear them.

Never bring contacts out in the field. Field conditions are often less sanitary, and it could easily lead to an eye infection.

Menstruation and Field Care

Practicing good field hygiene is essential for everyone. At basic training and beyond, you may need to be out in the field on short notice and for extended overnights, sometimes without access to supplies other than what you pack in your ruck or duffel.

When it comes to women's health, you should specifically consider the challenge of caring for yourself during your menstrual period. You can be thrown off your normal cycle when you are surrounded by other people who menstruate. It is a natural phenomenon that your body will sense other hormonal cycles and trigger your own. This happened to me. My period had just ended before I left for training and within a week of being at training, I started my period again. If you are on hormones or birth control pills, you may not experience this effect, since these help keep you on a set cycle. Physical and emotional stress can also contribute to menstrual cycle changes as well as potentially cause increased vaginal infections.

Pack extra pads and tampons with a range of absorbencies, and carry resealable plastic bags for disposal of used products. Stay as clean as possible. Wash your hands with soap and water when you can, but also pack hand sanitizer and scent-free wipes.

Stay hydrated and relieve yourself when you need to go. Unsanitary toilet facilities, lack of privacy, or the inconvenience of undressing when in full battle gear may tempt you to drink less water or hold your urine, but don't do that as it could lead to a urinary tract infection (UTI). A female urinary diversion device (FUDD) lets you urinate through your uniform fly while standing or sitting. Learning to use one can also help if you are in a situation that doesn't lend itself to a bathroom break or facilities. For example, if you deploy to a combat zone, you might be on long convoy trips and have to urinate within the vehicle to avoid improvised explosive devices.

Learning what works for you during basic training will help you later during field exercises, remote trainings, and deployments.

Hair Care

Decide the best way to minimize your hair care. While the military has authorized ponytails and braids once training is completed, many basic trainees must still meet hair standards of having a bun or short hair that does not extend past the collar. It is often up to leadership if you will fall under new standards or not.

My hair is straight and fine even without a blow dryer. I decided to cut my hair short enough that it never extended past my collar in training. In the morning, I quickly ran a brush through it. I did not have to worry about it being up in a bun. However, my friend has curly hair and a short cut would make her hair much more difficult to manage. Think about your hair type and figure out the best way for you to meet current regulations regarding hair care for your training. The goal when thinking about hair is to find a solution requiring as little of your energy as possible.

One other thing to note about hair is that the weather during your training can make a difference. Halfway through my training, I realized I was starting to get a rash because my hair never dried out each day from all the showers and sweating in the humidity of Florida in the summer. I began flipping over my head and running my fingers through it to get air moving through it before each night. It seemed to do the trick and I woke up with dry hair.

PRIVACY

Privacy at basic training is minimal. You will likely be in a large open bay with rows of bunk beds and wall lockers. There is not time to head to the bathroom and change in a stall. Showers are typically open bay, offering no privacy. The goal of basic training when it comes to general hygiene is to make it as quick as possible. Doors and other measures for privacy just add time.

But even with minimal privacy, it is important to pay attention to gender regulations for restrooms, showers, or bedrooms. Generally, the military keeps these areas single-gendered, with allowances for specific reasons. If something is going on that makes you worried for your safety, you should speak up about your concern to a staff member. It might be difficult to have that conversation, but your leaders are there not only to train you but to keep you safe as well.

11
AFTER BASIC TRAINING

YOU GRADUATED BASIC TRAINING! You are officially an airman, soldier, Coast Guardsman, Marine, sailor, or guardian. You might think that after thriving at basic training, the hard part is behind you. Well, not exactly.

There are often additional requirements, such as meeting specific physical fitness or mental stamina goals, for various career fields. Some technical schools are relatively easy, and the failure rate is low. Other schools are extremely difficult with a high failure rate. Just like basic training, not all members will finish the technical school they start. Even if you do not finish the required training, you are still a military member. You just will either have to start the training again or pick a new career field. Make sure you are prepared for your technical training, especially if it has physical fitness requirements. While basic training will help you get into better shape, the more in-shape you are before arrival, the less likely you are to be injured and more likely you are to be set up for success.

Lorraine talked about how she struggled in school to study and get good grades. She wanted to be an aircraft mechanic and did not want to be forced into another career field. She came up with a plan for how she would study and pass the tests for her training. She knew what was on the line and did not want the Navy to put her in another career field if she failed.

Technical training is a cross between basic training and full-time school. As training continues, you gain more freedom. Most often you still have formations and regular group exercises. You may or may not have weekends off. Even if you get time off, it may be limited to on-base activities. Near the end of training, you may be given a certain radius outside base to travel. Hair standards will follow normal regulations because you have passed basic training.

Once you graduate from your technical school, you will be assigned to a base or go back to your home station if you are in the Guard or Reserves. Some assignments are picked by class rank. So, if there is somewhere you really want to go, find out how assignments are given out and if your class rank is a determining factor. Either way, strive to do your best at your training to learn all you can and prepare for your career.

When you arrive at your new assignment, you will be required to live in the dorms unless you are married. Enlisted members normally live in the dorms until they reach the rank of E-4 or E-5. At my first base, members had their own rooms but shared a bathroom. The military is moving away from roommates and shared bathrooms in some branches. Even though dorms are primarily co-ed, you will only share a room or a bathroom with someone of the same gender. Living on base in the dorms gives you an opportunity to make friends. Rent is free and meals are provided by the dining facility three times a day. Many members do not need a car to get to work.

When you arrive at your first assignment, be ready for anything. Your unit could be getting ready to deploy, and you might find yourself on your first adventure before you get settled. Even if your first assignment does not start off with a rush of excitement, there will be a lot to learn. You spent months training to serve, and then weeks or months learning your career field. That information will help you as you begin your career. But it is just the beginning.

EMOTIONAL SUCCESS

BEING IN THE MILITARY IS MENTALLY CHALLENGING. Many military careers are unique to the military, without a civilian equivalent. Members often live in a location far from home, with the added challenges of deployments, trainings, and moving regularly.

The more you know about military life, the more prepared you can be for the challenges and the better choices you can make about your career field.

In this section, we will not only discuss challenges military members face in general, but how some challenges are more difficult for women. This section is meant to be a resource to help you make military life easier. If you find yourself in one of these situations or facing these challenges, know you are not alone. There are resources that can help you.

12
TRANSITIONS

CHANGE IS A CONSTANT in military life. Relocating from place to place, leaders and friends moving to new assignments, and the possibility of deployment or trainings keep a life in the military one of continual transition. While some people try to avoid change, the military can almost make change seem like a normal part of life. This does not mean change in the military is easy. But being prepared for it and learning how to cope with your emotions in healthy ways is important to help you deal with each change.

DEALING WITH EMOTIONS

When I think of emotions while in the military, I think about how the news hit me the day I found out I was deploying to Afghanistan with the Army. I had known I would deploy but when the orders came, it was as if all the fears I was keeping at bay about the dangers of deployment came to the surface. I cried hard. And just when I thought I had gained my composure, I cried again. New emotions kept surfacing as I worried about myself and having to tell my husband and parents the news. Eventually I told everyone, and it went way better than anticipated. I resolved to do my best and take it one day at a time. Sometimes I look back at that situation and wonder why I couldn't pull myself together. I knew it was coming, so why was the actual news so difficult?

I think I may have viewed my emotions as a weakness at the time because that is what the military teaches you. If you don't push past your emotions and react in certain wartime situations it can get you or your team members killed. That is why the military pushes you to be tough and not feel emotions. But emotions are a healthy part of the process of dealing with life events. And while not feeling emotions so that you can respond in a quick matter under extreme stress is important, avoiding emotions altogether is not healthy. So yes, I cried when I found out I was deploying. I thought I might die and that scared me. But I worked through my fears and acknowledged them as normal and valid. And because of that I was able to move forward with the deployment and go overseas and do my job. But before I could do any of that I had to process it.

You might not cry to deal with your emotions. But you need to find a healthy way to deal with and acknowledge your emotions. It is important to reflect on and acknowledge how you feel about the transitions in military life. The military may push you to put a smile on your face and just keep moving forward. But it is important to feel what you feel. Then when you have acknowledged it, find a way to move forward and get the mission done.

MAKING THE BEST OF RELOCATING

One really common transition in military life is moving. You will likely move a few times within the first year as you complete your basic training and career training and arrive at your first assignment. And then if you are on active duty, you will likely relocate to a new installation every two to three years. While sometimes moving so often can make it easier, I have found that the emotions that go along with moving are always challenging.

One way I try to make moving easier is by looking at each new assignment as an adventure and an opportunity. If given the option to drive to the next assignment, I create a road trip out of it, with various tourist stops along the way. My hardest move was after I got home from my deployment. I moved from New Mexico to Ohio to join my husband in the middle of winter. There was no fun road trip, and since I was still reintegrating back home from a deployment it was hard. I was also very lonely. I struggled to make friends at my office, and since it was cold and snowy, I didn't meet my neighbors or have the opportunity to build a community. When spring finally came, I started to get involved in different organizations and began building friendships.

Making friends at an assignment is a great way to make it easier to be in a new place. If there is an opportunity to join a club or participate in events you enjoy, you may be able to make friends quickly and make your new assignment feel like home. A military base is its own tiny city and, depending on how big the town or city near the base is, will give you varying levels of opportunities to get involved. At my first assignment there was an opening for the Society of Military Engineers group and all the lieutenants were dragged to attend with the commander. I decided to step up and run for one of the openings, and it gave me the chance to work with leaders and meet other members of the community. My friend loved softball and joined one of the softball teams on base. When her team needed an extra player, I joined her. It was great to meet others who worked on base and have fun. BBQs, movie nights, and get-togethers were also common events that provided the opportunity to spend time with friends or meet and make new ones.

At each location, take time to travel and see the places around you. Pretend you are a tourist and learn about what tourists come to see. My first assignment was near White Sands National Park, and I visited it a number of times and attended various special events. When I trained in Ohio for seven weeks, each weekend my friend and I drove in a different direction to see various sites. In Alabama, I learned more about the civil rights movement. There is so much history around the country. Take advantage of living in a new place and experience all you can.

Moving never gets easier. You get better at the process of moving as you do it more frequently, but each move is a new experience full of different emotions and challenges. Although I don't always look forward to each upcoming move, I always have a mindset that I will be moving soon. It is one way I mentally prepare for having to say goodbye and cope with having to start over again. And you never know, sometimes the military sends you to a place you end up loving so much that you decide to stay or move back one day.

PREPARING FOR A DEPLOYMENT

When you join the military, you should expect that you will go on a deployment. It does not matter if the country is at war or peace while you serve, the possibility of deploying is always present. Your likelihood of going on a deployment can vary based on outside factors, but it is a false assumption to think you will not deploy because there is no current conflict.

Like basic training, it is important to prepare mentally and physically for your deployment. A deployment is not the same experience as being at home station. You might find yourself being the only woman in your unit to deploy. You might find yourself in a dangerous location facing combat or other challenges. Often when members deploy, they spend a lot of time working. Depending on the mission, you may work long hours and the work may be physically exhausting, mentally draining, or both. When I was deployed, we worked every day. The mission of my unit required both office work and off-base missions. Every day was a new adventure but there were also a lot of challenges and always more work to do. Having work to do helped make the time pass but it also was a challenge because there were few days off. Even when I had a day off there was little to do besides work.

Technology often makes it easier to stay connected to loved ones back home during deployment. Most bases have Wi-Fi available. But being away from family and friends when you're deployed is different from when you are at home. You will likely miss out on special events, such as birthdays, anniversaries, weddings, and large family celebrations. It can be hard to not be there. Try to find out what options will be available and talk with your family about what communication may be like. Be creative about ways you can stay connected. Letters may seem old-fashioned but I cherish the letters I got while deployed and still have them today. I also hid notes all over my house that my husband found throughout the deployment. I did not have Wi-Fi so I set up a weekly meetup with my husband in the MWR tent so we could stay connected.

The military often puts more restrictions on its members while overseas, which may further change the dynamics of military life during a deployment. For example, depending on where you are located, alcohol could be prohibited, limited to a certain number of drinks, or only available at certain times.

Whatever emotions you feel about your deployment, it is okay to feel them. I felt excited, nervous, afraid, ill-prepared, unqualified, ready, and more before leaving for my training for Afghanistan. Having emotions is a normal part of the process. Feeling them and talking to a trusted friend, mentor, or counselor is one of the best ways to help you prepare for your deployment. Do not ignore how you feel. Even if you are excited, other emotions might be present. Working through these emotions before leaving for a deployment will help you in the long run.

Know that whatever assignment you receive, the military will train you to do what needs to be done to accomplish the mission. When I was in combat, I was amazed by how quickly my body was able to react and do exactly what I was trained for.

COMING HOME FROM A DEPLOYMENT

Just like leaving for a deployment is a transition, coming home is also a transition. You are not the same person that left for the deployment. How long you were gone and what you experienced play a role in how long it will take to adjust back to your new life. I unwisely believed the ten days the military gave me was all the time I would need to adjust back from being gone for a year. But I came back to a unit that had changed dramatically while I was deployed, my husband had moved to a new assignment, someone else was living in our old home, and I was working to prepare to move to my next assignment so I could be reunited with my husband. It is not surprising that life did not just go back to our previous normal.

Your life's dynamics will impact how quickly you adjust. For some people, it is a fairly seamless transition. They just pick up where they left off. Others take longer to adjust or need to reach out to someone to talk. Expect that reintegration will take time, and give yourself and your family grace to find a new normal. Children also have a range of experiences adjusting to Mom being home again. Some children pull away, others cling without letting Mom out of their sight. There is no set path for how you or your family members should handle the transition. Everyone has different past experiences and deployment experiences that can impact this transition.

Do not be afraid to reach out and get help either from mental health services or other organizations that can support you in the transition. Cohen Veterans Network offers free counseling in most states and is free for service members, veterans, and military family members. It can give you a place to go outside the chain of command. I knew there was something wrong when I came home. When I initially reached out for help, I was told I was fine. Instead of getting a second opinion, I pushed down the emotions and told myself I was fine. Years later, I realized my mental health was still suffering from my deployment. I was able to get help through both group and individual therapy. In hindsight, I wish I would have trusted my gut and continued to reach out for help after the first person told me I was fine. If you come home from a deployment and just feel off, don't doubt that feeling. Keep advocating for yourself so you can get the help you need.

Life after a deployment can be hard to adjust to but it is possible to find healing and balance with the experiences you had and your life back home.

SAYING GOODBYE

Being in the service requires you to say goodbye constantly. Often, it's your family and friends you say goodbye to when you head off to training or leave on a deployment. But they're not the only ones. Sometimes it is the friend you just met. Other times it is coworkers or training buddies who head off to a new assignment. Much of military service is a revolving door of new connections and subsequent goodbyes.

Saying goodbye to friends is hard. I always try to think back to the great memories made in the time we knew each other and hope that the friendship will continue even when we are separated. When saying goodbye to coworkers or friends, I like to instead say, "See you later." And while sometimes our paths won't cross again, life in the military is always changing and you never know when you might see a familiar face again. Social media can make it easier to stay connected to friends who move on to a new location. Most of my closest friends do not live in the city where I live. We stay connected through phone calls and text messages. Even with the distance, our friendship continues.

Saying goodbye to family is also hard. Maybe it will be the first time you will be living outside of your parents' home. It is okay to be sad; that is a normal emotion. But it is also exciting to be on your own and it is a necessary part of growing up. Again, technology makes it easier to stay connected to family even when far apart. And it is always great when you have the chance to go back home, or family can come visit in your new location.

As you get older and start your own family, saying goodbye for trainings or deployments is the hardest. But long-distance relationships are possible, and with technology, your loved ones are as close as a video chat. Make communicating with your spouse and kids a priority. At times you may not be able to talk as much as you want but do what you can to stay involved when you are away. And when you come back from being gone for an extended period, know it takes time to find your rhythm again. The military calls this time reintegration. How long you were gone and what you experienced while you were away can have an impact on how long the reintegration period will last. Take your time to find your new way forward.

When my husband and I were separated for more than a few weeks it always took a week or two for us to find our rhythm again. After being separated for fifteen months due to my deployment, it took over a month before we found a new normal. And as we added kids to our lives the dynamics continued to change. Reintegration takes time so be aware of the possibility of conflict and be open about how you are feeling and if you are struggling.

No matter what transition you face in military life, don't be afraid to feel your emotions. Good or bad, it is important to process how you are feeling. It is okay to be upset or disappointed. It is also great to be excited and optimistic. And it is even possible you will feel all the emotions I just listed and more at the same time. The military pushes you into situations that can be both exciting and scary. So, find a way to process those emotions and do not bottle them up. As we will talk about in the next chapter, stress is something you need to find healthy ways of dealing with. When stress is left unchecked, it can come out in negative ways. One of the first ways to start dealing with stress is feeling and processing your emotions.

13
STRESS MANAGEMENT

DEALING WITH STRESS in a healthy way is key to not getting overwhelmed. A study published by the US National Library of Medicine found that approximately 40 percent of service members (men and women) who participated in the study reported a great deal or fairly large amount of stress related to work. Interestingly, women responded that they had a higher level of stress at home and nearly a third of women reported a great deal of stress due to being a woman in the military.

Military life is stressful. No matter how many times I've packed up my house and moved to the next assignment, I always find new challenges I have to navigate. While the logistics may be similar each move, the emotions remain challenging. And moving, while very stressful, is personally on the lower end of stressors the military presents throughout a career. Living in a new location, being apart from family, deployments, temporary duty assignments, training exercises, training requirements, and loneliness are just a few of the many stressors that make military life challenging. Having constructive coping mechanisms is essential for dealing with stress.

POSITIVE SELF-CARE STRATEGIES

One of my favorite ways to deal with stress is working out. Some units require mandatory group physical fitness and others have a more laid-back approach. Besides needing to be prepared for the physical fitness exam military members must take, having a normal workout routine has many positive benefits for your health, mentally and physically. Find a way to prioritize physical fitness.

Amy Bushatz was struggling with her mental health. She decided to make an intentional point to go outside for at least twenty minutes every day. That was years ago, and it is a habit she has stuck to and highly recommends on her podcast *Humans Outside*. I never realized how many days I spent inside until I started following Amy's story. On days I feel my stress build, I now ask myself if I've been outside. Typically, my answer is "no." In a military environment, sometimes you will find yourself working in a windowless office or working odd shifts where you do not see the sun. Be mindful of how much time you are spending outside and how that affects your mood.

Mindfulness is another self-care strategy. If you are like me when I first heard about mindfulness, you might consider just skipping over this section. But my mom heard that mindfulness and meditation help people struggling with trauma from deployment, and I was desperate to try something to help. To my surprise, meditating every day helped me focus my thoughts and notice when I felt I was losing control. It is a daily practice that helps me, especially when life is challenging.

Being far away from family, working long hours, or doing a job you don't enjoy can create negative feelings for most people in the military (or in life in general). One great self-care strategy is to give yourself something to look forward to. It may be a fun vacation six months from now or it can even be something as simple as going to a movie or reading a book on your day off.

Lastly, if you are struggling, do not be afraid to reach out to someone at your mental health office or a base chaplain. You can also use resources off base, such as free mental health counseling with Cohen Veterans Network. Veterans Affairs centers may also be able to assist you. Going for mental health support is not a career killer. We all face challenges, and the military can make things harder than normal civilian life. My biggest regret is not going to get help sooner. Coping and getting by is not a way to live. Before getting help, it felt as though I was living in a dark room. I had been in the room for so long that I started seeing the world as normal through this dark view. Then when I reached out and got help, it was as if someone turned on the lights. My life completely changed. My mental health journey continues today as I am writing this book. I started mental health counseling due to the emotions I felt as the US withdrew from Afghanistan in 2021. Mental health care is a lifelong journey. It is always okay to seek support; it is vital to helping you get better.

NEGATIVE COPING STRATEGIES

Not all coping strategies move you forward. One common strategy that can work against you is isolation. If you are struggling, it may feel overwhelming to reach out and create a community of support; it might feel easier to hang out in your dorm alone and have limited contact with others. As an introvert myself, I recognize that having adequate alone time is important, but it is also important to not use time alone as a way to hide from the struggles you are experiencing. Make an effort to spend at least one night a week engaging with others, and work to build your community. As hard as it is to make that first friend, nurturing a supportive community can make the experience of military life better.

Extreme physical fitness or restrictive diets are another sign of negative coping habits. You may feel pressure to lose weight, especially as you prepare to get measured at upcoming physical fitness tests. Eating disorders in the military are a common struggle. While working out can be a great way to meet standards and improve your mental health, exercising to the point of exhaustion or using exercise as a way to punish yourself is not a healthy habit. It is also important to eat healthy, balanced meals to maintain overall wellness.

Alcohol is typically widely available, even to those not old enough to drink at a military base. Unfortunately, alcohol abuse is a common reason military members seek outpatient treatment. The military defines binge drinking as four or more drinks per sitting for women. If you regularly binge drink, you can build up a tolerance to alcohol requiring more alcohol to feel the effects, and this could lead to an addiction. Alcohol can also harm your mental aptitude. There are various rules in place regarding the time between last drinks and using heavy machinery or operating an aircraft. If you find yourself unable to control your urges, or you are drinking alone regularly or even during the workday, it is worth considering whether you might have an addiction.

Drugs can also be a negative coping method. The military has a strict drug use policy and does regular testing for its members through random drug tests. The rules for military members are often different from those in civilian sectors. For example, marijuana may be legal in the state you are stationed in, but any drug that is deemed illegal by the federal government is off-limits to all military members regardless of where they live.

POST-TRAUMATIC STRESS (PTS)

Stress in life is normal, especially in military life. But sometimes stress passes the level of "regular" stress and jumps to the level of trauma. This can be from war, military sexual trauma, a car accident, or other life circumstances. Do not make the mistake of believing that post-traumatic stress can only be related to combat or to a specific gender. Being in a war zone affects people differently, and PTS is triggered by more than incidents that happen in combat.

I personally struggle with trauma from my deployment and have utilized therapy to help deal with the challenges. While going through treatment, I learned that not all disorders from trauma are classified as post-traumatic stress. All levels of trauma are valid and getting help may be essential for moving forward.

For a long time, people believed women could not be affected by PTS in the military because women were not allowed in combat. But women have been on the front lines for years. Don't think you are immune from these diagnoses.

After my deployment, I had been back in the States for about two weeks when I felt something was off. I couldn't tell what it was, but I knew there was something wrong. I went to get help from a counselor on base and she told me it would probably take months before I felt back to normal. That conversation stopped me on my journey to find help and healing. Truthfully, I wasn't fine, but since I was told it would just take time, I accepted my life the way it was as my new normal. I finally got help, to deal with my anger and resentment, through a group recovery program six years after my deployment. Then in 2021, I began one-on-one counseling to deal with unresolved trauma and learn effective coping methods.

My story is an example of the stigma that women with PTS can sometimes face. The counselor I talked with did not take the time to ask me about my deployment experience, even though she knew I had been gone for a year. She didn't ask me what I was doing in Afghanistan. She did not know I faced a combat incident. When the counselor told me I would be fine, I believed her instead of following my gut that told me something was wrong. And because I didn't deal with it, the trauma came out in unhealthy behaviors (mainly anger) until I finally went and got help and began to find healing and move past the incidents.

I'm not the only one who experienced being dismissed when coming home from war. Jenny Pacanowski was running convoys as a medic in Iraq. She faced a number of combat incidents and many times did not think she would make it home. Jenny did not understand what was wrong when she came home from her deployment. It took her years to find help and deal with the trauma so she could find the right path forward after her deployment. Today she runs Women Veterans Empowered & Thriving.

Annette Whittenberger was able to function through her PTS from being assaulted when she was a cadet in Army ROTC until she deployed to Iraq. Seventeen years of stuffing down her feelings and then dealing with the loss and trauma of war led her to her breaking point. She said she was able to keep everything together, but then when she came home and had to deal with life's stressors, she felt overwhelmed. She was able to find help by sharing her story on her blog *A Wild Ride Called Life*.

FINDING HELP

If something happens to you and you feel you need mental health help, do not stop if one person says you are okay. You know how you feel and if there is something wrong, even if it is just a feeling, you should find out what is causing it.

I know from experience how hard it is to keep hitting dead ends and not be able to find help. But there are great organizations like Military OneSource, Cohen Veterans Network, and Veterans Affairs that can help veterans, service members, and their family members. Your local community may have additional resources that support women service members and veterans.

14
MENTAL TOUGHNESS

BEING A WOMAN IN THE MILITARY is challenging. Women make up less than 20 percent of the total force and, while this number is growing, women are likely to continue to be in the minority. This doesn't mean you can't do it or that you shouldn't join the military. But it is hard to be a woman in the military.

That is not to say military life is easy for men. Military life is hard for all service members, but women face unique challenges. It is important to be aware of these challenges as you make your choice to move forward in your military journey. The military is not something you can quit with two weeks' notice. Quitting is what the military calls Absent Without Official Leave (AWOL) and is punishable under the Uniform Code of Military Justice (UCMJ). So, if you are joining the military, know that it is a real commitment.

PRESSURE TO PROVE

One thing you will quickly learn when you join the military is how often you will have to prove yourself because of your gender. Unlike the men you will work with who may be given the benefit of the doubt, women regularly find themselves starting at an unfair disadvantage of having to fight a negative stereotype toward military women. You might have to work hard just to earn the respect given to the men in your unit. And because the military is a mobile force requiring you to move or to adjust to incoming new leaders, you might find yourself proving yourself over and over again.

Retired Brigadier General Wilma Vaught faced this issue in her career. She would arrive at a unit as a new commander and some of the men in her unit would try to leave, not wanting to work for a woman. As she was ready to move to the next assignment, those same men would share with her that while they originally didn't want to work for her because she was a woman, now they would follow her anywhere. She went into units at a disadvantage with men who did not want to work for a woman. But through her leadership and hard work she changed people's perceptions.

It might feel like you need to work twice as hard as the men you work with to prove you deserve to be there. But when you complete basic training and your career field training, you have earned your spot. The only thing you need to do is be competent in your job and work hard. People are watching what you do and waiting to see how you react. So, when you are given a task, do it at the highest level possible. All the military can ask for is your best. When you consistently provide high-quality work, people will notice.

Hana Romer felt the pressure to prove herself because she was a woman in the Marine Corps. Her job required hard manual labor and she pushed herself to do the same level of work as the men she worked with. She wanted to prove she deserved to be there. Looking back, she realized she did not need to prove anything. She had already done that by completing basic training and her career training school, just like her male counterparts. Instead of trying to show how tough she was, she wished she had only focused on being competent in her job and asked for help when needed instead of trying to do everything on her own.

It is never okay for you to be discriminated against because of your gender, race, religion, or sexuality. You should be held to the same standards as any other member of the military in your role. On some level, everyone has to prove themselves when they arrive at a new assignment. The men you work with still must prove their worth through the work they do. Every duty location has an Equal Opportunity Office. If you think you are being discriminated against, reach out to your office. They can help advise you on if you have a case. If you experience discrimination, keep a record of any incidents that take place so you can provide evidence, if needed.

In a male-dominated environment, you might feel pressure to ignore an uncomfortable joke, sexual comment, or other inappropriate action. If something in the workplace makes you feel uncomfortable, do not be afraid to speak up and address it.

A woman fighter pilot shared an experience that took place during her squadron's preflight brief. The briefer had included provocative photos of women in between each slide. At the end, her commander asked if anyone had a problem with the brief. No one spoke up. Then the commander explained that the pictures were not appropriate and he did not want his unit to be run in that manner. The woman fighter pilot had chosen to ignore the situation instead of speaking up when given the chance. She did not want to stand out and thought it would be easier to just go along with it even though it made her uncomfortable. The commander standing up and saying it was wrong made her realize how ignoring something made her not sensitive to it. She regretted not standing up and saying something.

Standing up takes a lot of courage and sometimes it is not received well. That is why it is important to build a tribe around you that you can trust to help bounce ideas off and create a safe place to blow off steam after a hard day. There is a time for action, and it is important to stand up for yourself. Knowing when to take a stand is crucial, and having the support of others will make it easier.

RUMOR MILL

Rumors and gossip seem to be a part of military life, especially when it comes to new women in the military work environment. Women are often a rarity in the military, so they attract attention.

How rare women are in a particular workplace seems to directly relate to the number of rumors circulating.

Work to stop the cycle of gossip. If you hear rumors about others, ask that the person telling you the rumor stop spreading rumors. And do not spread rumors about others yourself. It is easy to start talking about others and for stories to be embellished but spreading rumors can lead to hurt feelings. Also, do not be afraid to tell someone if you hear a rumor about them. I once heard a rumor about my friend. The person who told me made me promise not to talk to my friend about it. I stayed quiet. Months later she heard the rumor from someone else and confronted me, asking me if I knew what was being said about her. It had negative effects on our friendship, and I wish I had done what I knew to be right and told her.

If you hear a rumor about yourself, it may be best to ignore it. I heard a lot of untrue rumors about myself when I was deployed. I decided it was not worth trying to convince everyone else of what I knew to be true.

LONELINESS

Being a woman in the military can be very lonely. When I arrived at my first assignment, I received an email from a young lieutenant. She had been at the base for a few months and did not know any officers who were women. She found my name in an email chain and reached out. She was lonely and desperate for women friends.

The friendships and connections you build with fellow service members are important. Many of my friends during my service were men. But truthfully, there are some topics I could only talk to other women about. So, try to find women you can be friends with. Even if you are different and not best friends, having a woman you trust and can reach out to is invaluable.

Loneliness is often less of a problem at home station. I did not realize how lucky I was to deploy with another woman. Even though we were very different and may not have ever been friends outside of the deployment, we got along well. I regularly had a friend to talk to or hang out with while deployed. We are still friends today.

Air Force veteran Vanessa King had a different experience. It was hard for her to be the only female officer on her maintenance team in Afghanistan. Working the night shift meant she saw very few people outside her crew, and they were all men. She ran into a friend in the bathroom one night and they cried and hugged. It was only a few minutes of reconnecting, but it meant so much after feeling alone for so long.

Marine Corps veteran Katie Horgan's challenge with being a woman deployed was feeling like she could never let her guard down. As a Marine officer, she found herself constantly having to put up walls. It wasn't until she met up with a friend at a base she was passing through and they had lunch together that she felt she could lower her guard. It was hard to not have a safe person to confide in during her deployment.

Deployment is not the only time you might find yourself the only woman. When I went to a training with my Civil Engineering Squadron, I was the only woman from our unit of forty-some people. Our team met up with other units and there ended up being about a dozen women crammed into one tent for the training. On the last day of training, many teams left in the afternoon and headed home, but our flight was not until early the next morning. The final night at the camp all the women from the other teams had left, and I didn't feel safe in the tent alone. I ended up asking a senior noncommissioned officer if I could sleep in their tent. He was a father of two girls, and I had worked with him on a number of projects. I felt confident I could trust him. He understood my concern and told me I would be safe in their tent. Even though everything worked out, I was lucky. The person I depended on was trustworthy. Situations like this are hard and it is tricky to know what the right answer is. Trust your gut and do what you think is right.

YOU CAN DO IT

Being in the military is tough, but you are stronger than you know and can do more than you realize. Keep putting one foot in front of the other, and when you see an opportunity to connect with another woman, be brave and take the first step to say hello. The military changed my life for the better. It often pushed me to do things I never thought I could do. Take the positive aspects of the military and focus on them as you go through your military career.

15
SELF-PROTECTION

MILITARY SEXUAL TRAUMA (MST) is a very real issue. If you are a sexual trauma survivor, this chapter may trigger you in unexpected ways. The purpose of this chapter is to help protect you. Military members often work and live in common areas, including remote field locations, creating circumstances for potential rape or assault.

A major problem the military has faced around sexual assault and harassment was the power commanders had to overturn verdicts. Vanessa Guillén was murdered in 2020 after telling her harasser she was going to report him. It took national attention in her missing person case before her body was eventually found. Her story and many others led to the creation of the I Am Vanessa Guillén Act, which was passed in the 2021 National Defense Authorization Act (NDAA). The act changed the way the military responds to sexual harassment and sexual assault cases by making them stand-alone cases within the Uniform Code of Military Justice and moving the cases outside the chain of command. In the past, commanders could overturn a guilty verdict and erase all charges for sexual assault predators. This loophole has been closed. This act also gives survivors the ability to make claims for negligence and seek compensatory damages against DoD in cases of assault and rape. This landmark act has the potential to help change the culture of military sexual assault and harassment with policies and processes that focus on protecting survivors rather than perpetrators.

I have heard heart-breaking stories from women who were not able to get justice because the military mishandled the case or a commander repealed the guilty verdict. Others sought advice and, instead of getting support, were blamed for what happened. Experiences like these have led MST survivors to struggle for years with shame and trauma. I do not want this to happen to you. As much as the rules the military puts in place may feel like they are preventing you from enjoying your life, they are created to help keep you safe. Following the rules will not ensure you never face an assault or rape, but it can help decrease the chances of something happening.

Pay attention to and report negative behaviors that can lead to sexual assault, including lack of respect, gender discrimination, or sexual harassment. Do not tolerate any conduct that involves inappropriate touching, unwelcome sexual advances, requests for sexual favors, and deliberate or repeated offensive comments or gestures of a sexual nature. These factors and similar warning signs increase the likelihood of assault for both men and women.

If you are an MST survivor, get help. Talk to your Sexual Assault Response Coordinator to find out what options are available to you.

The Sexual Assault Prevention and Response Office (SAPRO) oversees the DoD sexual assault policy, working with each service branch and civilian communities to educate members about sexual assault and prevention and to support survivors. Visit SAPR.mil, and also review the sexual assault prevention program for your branch. Pay attention during the annual training to ensure you know what to do and how to help others if they face discrimination or assault.

PERSONAL SUCCESS

MONEY AND LOVE are two important topics to discuss when it comes to joining the military. For many military members, their military career is their first job. With it comes financial independence, so it is important to understand your pay and learn how to manage a budget. It is also critical to understand how saving for retirement while serving on active duty can have great benefits years down the road even if you only serve four years.

When it comes to relationships, life continues to get more complicated as you go from single to married. Adding children adds layers of responsibilities as well. The military is working to be more flexible for moms who serve. As you consider your professional choices for your military career, keep in mind that your personal finances and relationships also greatly impact your success.

16
FINANCIAL CHOICES

HOW YOU USE YOUR FIRST PAYCHECK with the military can impact your future. You might think your financial choices (good or bad) at the beginning of your working career do not matter, but they are critically important. Managing your money well in your first few years in the military can help set you up for financial success for the rest of your life.

There is more than one right way to set up your finances. Each person makes their own choices based on their unique circumstances. While this information is not professional financial advice nor comprehensive, it may help you consider how to avoid some of the common financial mistakes and challenges military members face.

COMMON FINANCIAL MISSTEPS

New enlisted or officer paychecks do not go very far. You can quickly spend your paycheck on going out to eat or buying various home items, and, if you buy a car, there are hidden expenses you may not expect. Besides your car payment, you will need to have auto insurance and be prepared for regular maintenance, repairs, and gas or electricity to get your vehicle everywhere you go.

If you do not need a car, consider not buying one. Many enlisted members live on base and can walk, ride a bike, or catch a ride from a friend to work. The longer you can go without a car, the longer you can build up your savings account and possibly save enough money to pay cash or put a sizable down payment on a car. If you do need a car, then it is great to consider a certified pre-owned vehicle. When I was in the Air Force, there was a used car lot on base that sometimes offered great deals on a car. No matter what you decide, do your research so you can be prepared for what to expect when buying a vehicle and all the additional expenses required.

If you are living in the dorms, even though dorm life may sometimes be frustrating, the positive aspect is that all your basic needs are met. You do not have to worry about utilities, rent, house maintenance, etc. You also can take advantage of free food at the dining facility for breakfast, lunch, and dinner. Having all those expenses covered helps stretch your paycheck.

When you do need to spend money, make good choices about what you buy. If you decide to make a big purchase, do your research first. You do not want to buy something and then realize you cannot afford it.

Credit cards or payday loans can cause financial strain or difficulty. You do not need a credit card when you join the military. The military will either give you a government credit card for travel or provide advance funding for travel. If you find yourself in a situation and need emergency funds, you can go to the finance office and get a no-interest loan. If you do decide to get a credit card, pay off the balance at the end of each billing cycle. The minimum amount due to the credit company each month may not even cover the interest that is added to your loan, thus causing a cycle of debt. If you get a credit card and are unable to pay off your balance each month, consider cutting up your credit card (or think of other ways to make it unavailable to you) so you can learn to live within your means without credit. Before I left for my deployment, my credit card information was stolen and the bank canceled my card. They sent a new one to my home address, but I was leaving the next day for Afghanistan. I had to deploy without a credit card. Truthfully, I was nervous about not having a credit card, but I never needed one. It might seem crazy to say you can live without a credit card, but it is possible.

CREATE A BUDGET

How do you get to the next payday (the first or the fifteenth) without running out of money? You create a budget. There are a lot of great resources for creating a budget, but most of them do not factor in the difference between civilian and military life.

A good rule of thumb when creating a budget is the 50-30-20 rule in civilian life. Fifty percent on needs, thirty percent on wants, and twenty percent in savings. Since your military paycheck may or may not include your Basic Allowance for Housing (BAH) or Basic Allowance for Subsistence (BAS) (money for food), it may be harder to calculate those percentages. So, when you are calculating your income, include the additional base pays you would receive (if you were getting BAH and BAS) to help you determine your total income and then each percentage of the 50-30-20 rule. BAH and BAS rates are available online. BAS is the same no matter where you live, while BAH is based on your rank and location.

Keep in mind that 50-30-20 is just a suggestion and based on where you live may not be feasible based on housing costs. You may choose to save more or add charitable giving into your percentages. Whatever you decide, make sure your percentages add up to 100 percent. The goal is to find a system that works for you so that you spend less money than you make.

Not everything in your budget is something you will spend money on every month. Unexpected expenses will arise. A budget doesn't mean at the end of each paycheck you are barely surviving to get to the last day before your bank account is refilled again. Instead, a budget gives you the peace of mind that you are prepared for unexpected expenses and you can plan and save for the things you really want.

Another great option people use for creating a budget is only using cash. When you go to a bar, restaurant, shopping, etc. and only have cash to make purchases, it makes it easy to stay within your means. If you bring a debit card or credit card, it is easier to buy one more item and you may overstep your budget.

There are also great online resources that help you track your finances. I use Mint to keep track of my family's budget. It is a great way to see where your money is going, and you can even set goals within Mint to help you stay on track as you are saving.

SAVE FOR YOUR FUTURE

Being young in the military is also a great time to build up your savings account and invest for your retirement. If you are single, all your needs might be covered by the military, which gives you the ability to save money at a higher rate than normal.

A great tip to help you boost your savings as you progress in service is to take all or half of the raise you receive when you get a pay raise and put it toward savings/retirement. If you are already living within your means, then hiding this money by putting it in a savings or retirement account can help you build up those accounts without having to cut anything out of your current budget. But also work to find the right balance. Saving money is great but it is also important to reward yourself for all the hard work you do. Finding the right balance between saving for the future and enjoying your life right now is important. Everyone has their own standard of living that makes them happy, so work to find that while also saving for the future.

BLENDED RETIREMENT SYSTEM

Now that you have a budget in place, let's look at where you put the money you are saving. Besides having a general savings account to save for vacation and unexpected expenses, you should also start saving for retirement. The new military retirement system is a great way to get started when it comes to planning for retirement. It does not matter if you serve one tour or twenty. Everyone can walk away with a great start to their retirement savings by understanding the new system and making the best financial choices while serving. The Blended Retirement System (BRS) was enacted on January 1, 2018. It replaces the legacy system of serving twenty years and drawing a pension based on your pay for the last three years of service. The new system has both a defined benefit plan and a defined contribution plan. To receive the defined benefit, you still must serve twenty years. But all members have the option to contribute to their Thrift Savings Plan with the military matching up to 5 percent of your base pay per year. The military will automatically contribute 1 percent of the service member's base pay after sixty days of military service, even if a military member does not contribute to their TSP. Military members are also automatically enrolled to contribute 3 percent to their TSP. You can stop this contribution at any time. After two years of service, the military will match the member's contribution up to an additional 4 percent, or a total of 5 percent. If you contribute 5 percent or higher, the military will match 5 percent. If you contribute 3 percent, the military will match that.

More than 80 percent of military members do not reach retirement eligibility. In the past, this left the majority of military members without any retirement benefit from the military. Now members can leave the military before the twenty-year point and have a jump start on their retirement. The best thing you have on your side when investing right now is time. Getting a head start by opening a retirement account right from the beginning of your military career can really impact your retirement no matter how long you serve.

TEMPORARY DUTY (TDY) AND PERMANENT CHANGE OF STATION (PCS)

When you travel for military work or training, or move to a new assignment, you get additional funds to help cover the costs. Make sure you understand these benefits so you do not end up overspending on your trips. While often members get per diem when they travel, sometimes if you stay on base and have access to the dining facility your per diem rate is decreased significantly. Knowing this will help you understand how much reimbursement to expect and may affect how and where you eat. Whatever you do, keep track of your finances when you travel so you don't find yourself not having enough money to cover the costs of your trip.

USE YOUR MILITARY BENEFITS

One great way to advance your financial plan is to use the educational benefits provided by the military. Starting your degree while you are serving on active duty using tuition assistance will help maximize your Post-9/11 GI Bill when you decide to get out. It can even make it possible for you to transfer the benefits over to your spouse or children.

PLAN FOR THE UNEXPECTED

You should start planning for your transition out of the military from the day you begin your military career. You never know what life will throw at you, and being prepared for an unexpected transition can make the transition easier and give you control over deciding to leave if an unexpected situation comes up.

If you do not have a financial safety net to help you as you prepare for transition, you may feel forced to stay in the military to continue to make ends meet. Debt can make the military feel like a must-do instead of what is best for you and your family. Many of the single mothers I talked to serve in the military until retirement because they know they are the sole provider, not only for themselves, but their children. Dr. Ann James talked about how important it was for her to have her finances squared away after her daughter was born. "Once I found out about my daughter, it changed everything for me, especially in regard to my finances," she said. "I made sure I had the focus and regained control on my finances."

Theresa Alexis faced a challenging situation when her daughter was born with medical issues. She got out of her contract early so she could take care of her daughter. She felt confident in this choice because she and her husband decided to live on one income. Even though her departure was unexpected, she was prepared. You never know what life will bring. Even with a contract, life can change quickly. Start saving and preparing for transition as soon as you can so you can be ready for whatever happens.

An injury or medical ailment could lead to a medical board causing you to have to leave the military earlier than you expected. A friend's husband recently faced this situation after having a seizure at work. The episode prompted a medical board decision for them to transition within months instead of three years later when he would meet retirement eligibility. It had a huge potential to throw a wrench in everything they were planning for in the future.

An assignment change, deployment, or unexpected move might be another reason you find yourself leaving the military earlier than expected. The military can require members to move to places they do not want to move to. Depending on how close to the end of a service commitment they are, they may decide to say no to the assignment and instead leave the military. My friend's husband was close to the point to stay in or get out. The next assignment was a year alone in Korea. Even though they had planned to continue serving, they did not want to spend a year separated and instead decided to leave the military. He had his degree and they did not have debt, so they were able to make this choice and quickly start planning for the next phase of their life. Being financially stable gave them options.

Not everything that's unexpected means an early end to your military career. The military (and life in general) is full of surprises. For instance, depending on where you are stationed, you may deal with unexpected emergencies due to weather, forest fires, earthquakes, flooding, tornadoes, extreme cold, and more. Living in a different part of the country, you may be exposed to all kinds of conditions. I grew up in California, where the weather is typically mild but there are also earthquakes. When I went to Florida for my basic training, I had never seen lighting as intense as the storms I saw there. I learned a lot about how different and dangerous the weather is. Living in multiple places across the country has taught me that each geographic area has its challenges.

In addition to weather and natural disasters, wildlife dangers change depending on where you are located. In some areas, ticks and Lyme disease are cause for concern. In other parts of the country, there are dangerous snakes or spiders that you might need to watch out for. And if you live in Florida or other parts of the South, you may need to be aware of the danger of crocodiles and alligators.

Become an expert on your environment. Find out if there are any safety precautions you need to be aware of at each new assignment so you can be better prepared for the unexpected.

No matter what situations you find yourself in, being intentional about your finances, managing your spending and savings, limiting debt, and saving for retirement early on will put you in the best position to support yourself and keep your options open now and in the future.

17
LOVE AND RELATIONSHIPS

RELATIONSHIPS IN THE MILITARY ARE CHALLENGING. Know the values you want to live by and strive to honor your values and respect others'. Military life is hard on friendships and romantic relationships, with civilians as well as with other service members. As a service member, you face a tremendous amount of stress and uncertainty that may impact the emotions you bring to a relationship. Your friends or partners may be directly impacted by military life stress as your relationship becomes more serious and they go through your absence during a deployment or move with you to your next assignment. There are many aspects of relationships to consider, including being true to yourself.

SINGLE IN THE MILITARY

If you are single, enjoy your independence! Don't rush to be in a relationship. There are many programs available to single service members in which you can enjoy friendships doing activities with others who share your interests. Visit your local morale office for special tickets and excursions especially for single members. This is a great time in your life to keep your possessions and expenses light, your belongings easily transportable, and simply enjoy your service lifestyle in your local community. There will be plenty of time later to engage in more serious relationships. Being single allows you to focus on your career, set yourself up to meet your goals, and build a community of friends for a lifetime.

If you feel ready for a romantic relationship, know yourself and what you want out of a relationship. When I met my husband, I was happy and confident with who I was. I even avoided dating him at first because I did not want to be in a relationship. Eventually, our friendship grew into love. I am so thankful I had taken the time to really figure out my own sense of self before I met my husband.

DATING IN THE MILITARY

Several veterans contributed to the following list of "rules" they would tell their younger selves about dating. Customize them to fit your lifestyle and personality.

- ★ Explore your career options, travel, make friends, and really get to know and appreciate yourself as a single person. Do what is best for you; don't make choices based on what others want for your life.

- ★ Do not date anyone for at least the first six months at an assignment. Take your time getting to know the base and the people you meet before committing to a relationship. Make friends and spend time in group settings.

- ★ If you meet someone who is unwilling to respect your boundaries and the choices you have made for yourself, do not dismiss this view of their character.

- ★ Know your worth.

- ★ Be happy with who you are before you start dating. A person cannot fill a void inside you. Building a relationship on a lie or pretending to be someone you are not never works out well.

- If you can't hang out in a group setting, meet in a public place. Always tell at least one friend where you are going and who you are meeting.

- Lock your doors.

- If you drink alcohol, have at least one other trusted person with you who is not drinking.

- When you start dating, take it slow. There is no hurry.

- Dating within your unit is generally discouraged.

- Relationships in the military require skills for communicating and navigating challenges together in person as well as during long periods of separation. A relationship that works well during deployment may or may not work well when together, and vice versa.

No matter what rules you adopt or create for yourself, be intentional about nurturing the relationships you want at this time in your life. Surrounding yourself with quality, meaningful friendships will provide a supportive community as you continue on your service journey and will likely have a positive effect on everything you do, including any future romantic relationships.

FRATERNIZATION

The military has rules for just about everything, including who you can date. The goal of fraternization policies is to help define acceptable and unacceptable relationships between colleagues. These rules aren't intended to keep people from having personal relationships or team building. Instead, they are meant to help prevent the unfair treatment or appearance of unfair treatment between supervisor and subordinate relationships.

In general, officers and enlisted members cannot date, and military members are not allowed to date members in their chain of command. It is recommended that members within the same unit do not date. And it is also frowned upon for younger ranked members to date senior ranking members. Because the people you work with in the military are often who you spend your time with outside the office, the lines of fraternization can become an issue.

Fraternization rules do not stop some service members from using their authority as a means to coerce you into dating them or performing sexual favors. This is NOT OKAY. If something like this happens, you should report it to your Military Sexual Trauma advisor so that it can be dealt with. It may seem dramatic to get advice from the MST representative, but they will confidentially help you determine the best way to handle your situation.

MARRIED IN THE MILITARY

Being married is hard in general. Being married in the military is even more challenging. I have been married for more than fifteen years. All that time my husband has been in the military. I can't imagine my life without my husband, but we still faced many challenges. I was once asked, "How did you stay married when you were separated from your husband for over a year while you deployed?" The answer comes down to communication. When I deployed, there was no Wi-Fi on my Forward Operating Base and I did not have a cell phone while overseas. But we still communicated regularly. I still have all the emails we wrote during the deployment. In them, we talked about the basic logistics of coordinating various things, along with personal updates. I also wrote him handwritten letters. He sent me care packages with short notes. We also were able to have weekly video chats in the MWR tent. By the time I left Afghanistan, the main base I went to before flying home had Wi-Fi and it was great to be able to talk to my husband and parents in my bunk instead of having to wait for a computer in the MWR tent.

Another woman who deployed to Afghanistan was able to attend her daughter's parent-teacher conferences and video chat with her daughter almost every day. Even with advanced technology, it is still hard and takes effort.

Communication continues to be important for our marriage. My husband still travels for work, and we face separation. It is also how my children stay connected to their dad when he travels.

In addition to communication, you will need a lot of patience, adaptability, and the willingness to work as a team. There will likely be many situations in which one or the other partner will do more than their share, especially during times of separation, adjustments to new assignments, and if you become parents. Give 100 percent to your relationship, and don't keep score. Together you can face whatever military life brings your way.

YOUNG AND IN LOVE

Some military members get married very young. Getting married to get stationed together is rarely a good option. Military life will require time apart regardless, so it is better to learn if the relationship can work long distance before jumping into a marriage.

If you are both service members, being married does not guarantee the military will station you together. My husband graduated a year before me. We were separated between his graduation and when I went active duty a year later. I only saw him at the holidays and my graduation. My first assignment on active duty was in Alabama for six weeks of training before finally moving to my husband's base. While we were both on active duty, we spent approximately three of the six years I served apart due to deployments and trainings. Being separated is hard but if you create a strong foundation built on communication your relationship can last.

Sometimes people may consider marriage as a way to escape the dorms. While getting married and having the opportunity to live off base with your significant other might seem ideal, there are advantages to living in the dorms. Having three meals provided for you at the dining facility every day is one such advantage. Another is having a place to live that has furniture you don't have to pay for. The dorms also have the convenience of being close to work and your fellow military members. Getting out of the dorms is not a reason to get married. Sooner than you know you will make rank and have the option to not live in the dorms.

Eventually you may decide you want to marry someone you met in the service. Whirlwind romances and short engagements do not equal a failed marriage. I recommend taking your time but in no way am I saying getting married fast does not lead to a lasting relationship. Just make sure you have thought it through and are getting married because it is what you want and not the result of circumstances influenced by the military. Rushing into marriage with the wrong person for the wrong reasons is not a good solution. But marrying someone you love can make the military experience better. It will be challenging but it also can be very rewarding.

When you get married to another member of the military it is commonly referred to as mil to mil or dual military. Almost half of married service women are married to a service member. There are several advantages to both members being in the military, especially before having children. My husband and I were dual military for six years. It was great for us to both have full-time jobs wherever we were stationed. It made moving easier (but we never actually moved at the same time) because we each moved with a career instead of one of us having to start over. But even with the advantages, it does not change the fact it is difficult and requires a lot of sacrifice. Did you see where I mentioned we never moved to a new duty station at the same time? That is only the beginning of the challenges faced by military members who are married to each other.

Having a spouse who is not in the military can make it easier to stay together but military spouses often face challenges too and are required to make sacrifices of their own. Long hours, deployments, and trainings are a regular part of military service. Moves can be frequent and unexpected; being far away from family can put a heavy strain on military marriages. Military spouses often struggle to continue a career at new assignments. Military spouse unemployment runs 18-24 percent, about four to five times higher than general civilian unemployment.

Be upfront and honest with your significant other, especially if you plan to continue serving past your first commitment.

18
MOTHERHOOD

IF YOU WANT TO BE A MOM while serving in the military, it is possible. The military has made a number of accommodations to make it easier to be pregnant and to be a mother in the service.

One of the key reasons I left the military was because I knew that approximately six months after giving birth to my son I would likely deploy. Deployments for new moms now cannot happen within the first year of a child's life. Moms still have the option to choose to deploy before this window closes but now there is protection to allow moms to be home for the first year.

Another recent change is the amount of time moms are given to recover after giving birth. Previously, moms were allowed six to eight weeks of maternity leave depending on doctor recommendations. New moms now get twelve weeks. This allows moms more time to heal and connect with their babies.

Even with the positive changes, women still leave the military at a higher rate than men. Quality childcare or family planning are major factors in their choice to leave. Regular moves also factor in. It is one thing to be a working mom with a community to support you; military moms have to rebuild that community every few years with each move.

But it is not impossible to serve in the military and be a mom. So do not be discouraged. There are resources that can help you, and the more aware you are of the challenges, the better you can prepare for motherhood. Air Force colonel Rojan Robotham wrote *Working Moms: How We Do It* to encourage women to continue their careers in the military after having children.

Another great option for women is transitioning to the Reserves or National Guard. This option can give you the balance of family life and more control over where you live while still continuing to serve in the military. The Reserves and National Guard can also be a temporary assignment giving you the option to go back to active duty as your kids get older or life circumstances change.

SUPPORT SYSTEMS

All families need support systems, and this is especially true of moms who serve. The demands of military service continue even when your child is sick or is celebrating a birthday or is struggling with school work or wants you to attend their soccer game.

Retired Brigadier General Carol Eggert asked her children if they wished she had been home with them more instead of serving in the military. "And they all said, 'No, Mom, it's way cooler to have a mom in the military.'" Listening to her kids' perspective made her feel less guilty about leaving for deployments and trainings so often while they were growing up. She served in the National Guard and was able to go in and out of active duty with the support of her husband. She said she was able to do what she did because of their strong marriage. Her husband did not serve in the military and was able to provide stability and support at home and be more involved in parenting. Having a supportive spouse can make balancing parenthood and the military easier.

But not all moms in the military have a spouse to rely on. Single moms continue to make a huge impact in the military. Single moms rely on a support network to get through the challenges of serving. Some rely on extended family to watch their children during deployments. Others have friends who are like family who fill that role. Naomi Mercer was a single mom facing a deployment. Her ex-husband was also deployed so he could not take care of their daughter. She relied on her long-term care provider, her best friend. Her daughter still talks fondly of the year she spent with her friend's family. Naomi was able to focus on her mission while relying on a friend to take care of her family while she was deployed.

Mona Johnson joined the Army as a nurse with two young girls. She relied on her parents to watch her children when she went to various trainings. While at her first assignment, she primarily worked the night shift and had to rely on babysitters who could stay the night. Luckily, after her first assignment, she worked more traditional hours, which made it easier to find reliable childcare.

All moms face a lot of pressure, but it seems single moms face an extra measure of it. Not only do they have to take care of their children, but they are also often the sole financial provider. Sometimes they decide to stay in the military because of the benefits of a steady paycheck and medical insurance.

DEPLOYING AS A MOM

Deploying or attending extended training is a regular part of military service. Many women who have deployed and had to leave their children behind talk about how difficult it is to say goodbye. Cynthia Cline shared what it was like to look at her thirteen-month-old daughter before leaving for a six-month deployment. She said, "I tried to memorize what it was like to nurse her one last time. Desperate to freeze that night for as long as I could, I held on tight, refusing to let my tears flow." Moms in the military are called on to make hard sacrifices, especially when a deployment means they miss a part of their children's lives.

There are deployment resources on base to help you prepare for being apart from your children, and to help support your children while you are gone. Talk with your child's teachers, coaches, and others who will be caring for them in any capacity. This is the time to call in your village! Each child, depending on their age and personality, will respond uniquely to your deployment. While technology may help you stay connected from afar, you want as much support as possible for your family, in your home and community.

STAYING OR LEAVING

When it comes time for you to decide whether to stay or leave the military, take the time to look at the pros and cons to help you decide the best path forward for you. Think about the sacrifices that may be required if you continue to serve and if you are willing to make those sacrifices. But also think of the opportunities you may have if you stay in the military and the example you can be for your children. Sometimes, you might be at a point in your commitment that leaving the military right away is not an option. Use the time you have left in the military to learn as much as you can to help you determine if you want to continue serving or if you feel called to a different career.

You are the only one who can determine what path you take. Many women continue to serve and are able to balance the challenges they face as both moms and service members. Others choose to leave. There is no wrong choice; you just have to decide what is best for you and your family.

Questions to Consider:

- ★ Do you want to continue to serve?
- ★ Do you have a good support network?
- ★ What challenges might staying in the military create? Are you willing to continue serving with those challenges?
- ★ What are your goals related to military service and your family? Is there a way to achieve both?

CAREER SUCCESS

HAVING A CAREER IN THE MILITARY will open your world through the work you do and the places you travel. Learning a new skill and earning your veteran benefits can help you for the rest of your life. There is a difference between having a career and thriving in a career and using it to set yourself up for success. Through these chapters you will learn tools that will help you prepare for future career success.

Everyone transitions out of the military at some point and many members only serve one tour of duty before they move on. By finding mentors, saying yes to the right opportunities, and preparing for life after the military from early in your career, you position yourself for success both while you serve and when you leave the military behind.

19
MENTORS

ONE OF THE FIRST LESSONS basic training teaches you is that you cannot do it alone. And while being part of a team is essential for getting through and thriving at basic training, it also is a tool you can continue to use throughout your career. It is difficult to succeed in the military without mentors to help you make choices and encourage you to pursue opportunities you would not have without their recommendations.

One of my biggest mistakes in my career was not seeking out advice from leaders who could have given me a new perspective. My second commander was married to a military service member and I did not take the time to ask him and his wife about their experience. I missed out on an opportunity to learn from them. I often wonder if it would have changed my choice to get out of the military. Instead, I regularly tried to do it on my own instead of relying on the leaders around me who had more experience and were willing to help me meet the dreams I was working toward.

If you want to do something or are given an opportunity, talk to your supervisor or other senior members you trust. The more people you talk to, the greater the depth of advice you will receive, and someone may have a connection that can help you achieve your goal. Mentorship does not have to be a formal one-on-one monthly meeting. With military members moving and leadership changing, it can be hard to have a mentorship relationship with one person. Instead, mentorship can be looking to the leaders you have around you at the time and using their advice to help you as life situations come up. As your network grows you can always reach back to those you worked with in the past.

Be aware that leaders are watching the work you do, and often opportunities like assignments, early promotion, and leadership positions happen because leaders see your commitment and work ethic. In *A Higher Standard*, General Ann Dunwoody talks about how often she relied on mentors and how often leaders advocated for her without her knowledge. General Dunwoody was the first woman in the Army to achieve the rank of four-star general. She said she never would have achieved that rank without the leaders who believed in her and advocated for her for different opportunities. Some of this was luck but she also did her best in every job she had. People noticed her dedication and skill.

Not everyone will achieve the rank of four-star general. But if you show your dedication and a willingness to learn, it can open doors to opportunities you would not expect.

FINDING A MENTOR

According to *Harvard Business Review*, 76 percent of people say mentors are important, but only 37 percent actually have one. They attribute this gap in large part to the fear of rejection. People are afraid to ask for that initial meeting. This might be something you experience as you look for a mentor in your military career. There may be someone you admire that you would like to gain advice from, but you are too afraid to ask for help. A no answer is not a failure! You have many options for mentors. If your ideal potential mentor says no, don't take it personally. There are many reasons why being a mentor may not be a good fit for them at the time. In that case, you can ask if they have a suggestion for someone else who could provide mentorship.

Perhaps you are hesitant because you aren't sure how to best engage with a mentor. The first step to establishing a good mentor relationship is figuring out what you hope to gain from the mentorship so you can match your goals with a mentor who has experience in that area.

If you are planning to transition out of the military at the end of your service commitment, finding a senior leader in the military is probably not the best option for you. Instead, you should try to find someone who has transitioned out of the military, ideally in the civilian career field you want to transition into. Resources like Veterati, a free service that provides one-hour mentorship phone calls with successful professionals, can help you connect with someone and receive sound advice. That phone call could turn into a more formal mentor relationship.

If your goal is to serve in the military for twenty years and reach a certain rank or become certified in a certain field, look for someone who has done that or is on the path toward that goal.

Be open to mentors from different backgrounds than you. Differences in gender, race, ethnicity, religion, political views, education, upbringing, and life experiences can all add depth to your conversations and help you view various problems from a perspective you may not have thought of on your own.

And remember that to be a good mentee, you need to drive the relationship from the very beginning.

WHAT MAKES A GOOD MENTEE?

Mentorship starts with the mentee reaching out for advice. Leaders may advocate for you and help you gain opportunities but to learn from them and grow as a leader yourself requires you to step up and ask for the support you want.

To be a good mentee you must be open to feedback. Constructive criticism is essential to good mentorship. Learning the skill of being able to take constructive criticism and using it to change yourself for the better is not a natural skill. When I first joined the military, I struggled when I received negative feedback. I saw these negative traits in my character as deep flaws. At the time, my perfectionist personality made it hard for me to receive the feedback. Instead of using it as a tool, I ignored it. All feedback, positive and negative, is valuable. Be open to hearing feedback and then take small steps forward to be the person your mentor knows you can be and most importantly, you want to be.

Another important trait of mentees is that they are prepared and do not waste their mentor's time. Be prepared with specific questions for your mentor. Leaders in the military are very busy but all my commanders had an open-door policy and wanted to help mentor their officers. Good leaders make time for those who need their help, but you must make sure to respect their time.

Military members often skip small talk and jump straight to the point. Knowing this about military leaders will help you prepare for the meeting. Know what your problem is and how they can help and tell them up front what you are looking for. This will help you get the advice you are looking for and not waste time. And do not be surprised if they need time before they can give you an answer. A quick answer is not always the best one. If you have limited time to make a decision, make sure they know that as well so they can help you get the answer you need before you have to make your decision.

Not all mentor relationships require formal meetings and a strategy to grow. Many mentor relationships are informal and may come up based on where you work, a particular problem you are facing, or certain life challenges. Take advantage of getting feedback and support when offered by others.

PEER REVIEWS

Military leaders occasionally do peer evaluations at the end of trainings or exercises to help improve both the training and you. During these evaluations, you and your peers are instructed to provide one negative and one positive thing about each member of the team. And while it will likely be easy to write down positives about your team members, it may not be so easy to write down something negative. We all have things we need to work on to become better leaders and teammates. Don't let these evaluations defeat you.

When I first received a peer evaluation, I was so overwhelmed by the negative feedback that I did not even read the positive feedback. I read each negative as a direct criticism, not because they were not true, but because I thought I had hidden my faults from others. Knowing that people could see the areas I struggled with was difficult. Years later I stumbled upon a box of papers from my time in the military. I found the feedback and remembered the negative feelings I had. For some reason I decided to read it again. I was shocked. It was overall positive, and the nice things people said were so encouraging to me. The negative feedback I could originally not even finish reading was actually focused on personality traits I had been working on over the years since leaving the military in my own personal journey. I realized I missed an important lesson in receiving this feedback. Feedback helps you grow and helps you become a better person.

But this does not mean if you receive a peer review that you have to follow every piece of advice in it. Some of the negative feedback I received was about being shy and introverted. And I am both of those things. You do not have to change who you are to be a good service member. Each person brings their own unique gifts to the table. Have I learned to push past my shyness and step out in front of a crowd? Yes. But it will not be my natural instinct to jump in and be out front. Be who you are and be proud of that. When it comes to constructive criticism you cannot make everyone happy. We all view the world in our own unique way. Take what you can and work to be better and do your best.

20
CAREER ADVANCEMENT AND PROMOTION

MAKING RANK OR PROMOTION are terms that mean advancing in your career in the military. For the first few years of your military career, rank advancement is determined by time in service. As you increase in rank, promotion begins to be based on how you perform in the military and on tests, and on whether you obtain certain certifications.

One thing to note is that service members do not all enter with the same rank. When you first join the military, your rank is determined by various factors. These are typically check-the-box classifications. A common example for enlisted members is if you have at least sixty college credits you start at the rank of E-3. To become an officer, you have to complete a specialized program for officers. Officers generally start as an O-1, but some career fields, like doctors, lawyers, or dentists, start at a higher rank.

As you progress in rank it becomes more challenging to be promoted. At different points, based on your branch and career path, you will need different qualifications. A standard measurement for all service members is a performance report. Everyone receives a yearly report that includes an overview of the work they did for the military and some level of ranking or meeting standards evaluation. Your physical fitness assessment may also be included in this report either by number score or a pass-fail metric.

Some promotions require members to test to show their knowledge for their career field and branch. Other times certain training is required to advance in rank. While your supervisor should help you prepare for each stage of the next rank, it is your responsibility to ensure all the information you have is correct and that you advocate for yourself. If your performance report has negative feedback and you did not receive counseling, you have the right to try and correct your record. You are your best advocate, so do not be afraid to ask for feedback and work to get your records corrected if you do not feel you are being treated fairly. Also know that one bad performance report does not mean the end of your career. It may require more time to promote to the next rank but often you can overcome the challenges. Just keep doing your best and work to improve.

Sometimes the limitations put on pregnant women can be a particular challenge. Many military trainings are not open to pregnant women, including basic training. If you sign up to enlist and then become pregnant before basic training begins you have two options. You can choose not to join the military, or you can plan to attend training after you give birth. But this is just the beginning of training limitations. Other military trainings that may be required for career promotion or to be in command or leadership positions can be limited if you are pregnant. Based on your career path, not going to training because of pregnancy could have a negative impact on whether you get promoted and could limit your options in the future. Sometimes a pregnancy is unplanned, but if you are interested both in promotion and in starting your family, looking at these training requirements can help you plan for the best time to start your family.

The military has other restrictions for pregnant women as

well. Pregnant women cannot deploy and if they become pregnant during a deployment, they will be sent home early. There also can be job restrictions around pregnancy depending on your career field. For example, women cannot be aboard ships in the Navy past the twenty-week mark and most often women pilots are restricted from flying at twenty weeks as well. There are a number of other career fields that may have limitations put in place to protect both mothers and their babies.

The limitations around pregnancy and childbirth are not supposed to impact your career but unfortunately some women have faced discrimination or have found challenges while being pregnant or after giving birth.

CROSS-TRAINING

If you are in a career field you do not like or if you are interested in doing something new, then cross-training into a new career field is often an option. Some members even cross-train into different career fields because they have a greater opportunity to get promoted. Keep in mind that cross-training may require attending school for that career field and could lead to an additional service commitment.

21
EXIT STRATEGY

EVENTUALLY, EVERYONE LEAVES the military. Discovering the best time to leave service is often not simple. Even those who serve until retirement must decide when they will take off their uniform. There is honor in your service, no matter how long you serve. Many veterans express feeling like they did not do enough while they were serving. For a long time, I regretted leaving the military after six years of service even though it was the right choice for me and my family. If you're thinking about leaving the military and feel that you have not done enough, know you are not alone in that feeling and ask yourself what more you could do. The answer to that question could help you decide if it is the right time to leave. That feeling of wanting to continue to give back is part of who you are. Serving in the military is just one way to give back to your community.

DECIDING YOUR NEXT STEP

If you are struggling with what your next step will be, think back to when you first decided to join the military. Remember your reasons? Have you accomplished what you hoped to achieve by serving in the military? If you wanted to find a way to pay for college, and you are now eligible to have your tuition paid and feel you are ready to go back to school, then leaving may be the best option.

Also consider how much you have changed since joining the military. You may have new goals you'd like to accomplish, either in the military or beyond your service, and maybe the timing is not right yet to make any moves.

Consider your financial stability. If you are in debt and need a secure job, staying in the military might be the best option. Or perhaps the civilian equivalent of your job pays much better and you have an opportunity waiting for you if you leave.

Sometimes it is tricky to figure out the best way forward and you just have to make the best choice you can. I didn't feel ready to leave the military, but I also knew it was the best choice for me and my family. It was not easy, but it was the right choice.

Other times the choice to stay in or leave is clear. Maybe you are ready for the freedom to make your own choices. Or you are in a job you do not enjoy and the military is not allowing you to switch career fields. There are too many potential situations to include here. Whatever your situation, there are likely pros and cons to leaving or staying. When your reasons to leave outweigh your reasons to stay, it's time to start preparing to leave the military and figuring out your next step.

Don't forget you also have the option of leaving active duty but continuing service through the National Guard or Reserves.

FEELING GUILTY

Once you make your decision, do not feel guilty either way. Some people may question your decision. They may not understand why you want to leave or continue your military career. This is your choice. Do what is best for you and your immediate family.

That does not mean you should not ask for advice from leadership or peers you trust. Getting advice from others is a great way to sort out how you feel and help you decide if it is a good time to leave. Good leaders want to see you succeed both in and out of the military and will help you find the next best step for you. But the final choice is always yours when deciding if it is the right time to leave military service behind.

LEAVING THE MILITARY

When you leave the military, you become a veteran. The veteran community is an incredibly valuable resource. Your military service opens doors to opportunities unavailable to those who have not served.

The veteran community is diverse; it's a mix of all branches, ranks, and careers. There is a special bond among veterans that creates an instant connection. Many veterans love helping other veterans as they start their journeys after military life.

When I left the military, I felt like an outsider at first. I did not want to get involved in the veteran community because I believed I would not be welcomed because of my gender. But I have found the veteran community to be very welcoming. I am connected with other veterans on LinkedIn and have attended many networking events with other veterans. Even when I do not know anyone at these events I can connect with others through my service. The veteran community is full of amazing people who want to help see their brothers and sisters in service succeed.

There are many different organizations that provide career mentorship to veterans and help them find jobs or start a business. Veterans Affairs is just the beginning when it comes to veteran benefits and support. There are also veteran organizations just for women. I love being a part of Women Veterans Alliance and Women Veterans Interactive. These groups have yearly conferences where women veterans can connect with each other while learning about available resources. We need the voices of young women who have served to join these organizations and help the veteran community continue to thrive.

VA BENEFITS

When you prepare to leave the military, ensure the VA has all your correct information so you can use your benefits. If you are planning to go back to school, make sure everything is set up so you can use your earned GI Bill benefits.

It's also important for you to go through your medical record very carefully to make sure you have all your information and that it is current and accurate. You'll want to file a medical claim with the VA. And you should review your file with a claim advisor before you leave the military. They may not discover any medical claims but if they do you will potentially be eligible for medical disability benefits. You will also qualify for care for any issues they find during the physical through the VA.

Even if you do not have any medical issues, you should register with the VA and complete your yearly medical physical so that if underlining health conditions from your service arise you will already be part of the VA system. Even if you have health care and do not think you will ever need VA services, make sure to register. I did not get this advice and wish I had known how important it was to register and file a claim when leaving active duty. If you do not register within five years of separating from the military, it makes it much harder to file a claim with the VA if medical issues arise.

FINDING A JOB

One of the hardest parts of leaving the military is finding the right job for you. Many military members rush into a job because of the security of a paycheck instead of taking their time to learn about what they want to do and find the right company for them. I love the book *B.R.A.N.D. Before Your Resumé* written by Air Force veteran Graciela Tiscareño-Sato. Graciela writes about her journey to leave the military and the process she went through to discover her next step. She was a navigator and there was not a job that closely matched the skills she had gained from the military. Through mentorship she was able to get a job focused on her skill as a marketer. Eventually she started her own company writing and speaking.

Other programs such as Veterati can connect you with a mentor to help you as you figure out your next step. Veterati offers free one-hour mentorship blocks with veterans. Vets2Industry hosts quarterly networking events to connect transitioning veterans with companies looking to hire veterans. Military Talent Partners helps connect recruiters with veterans to help them find their next career opportunity. There are a number of other resources that can help you in the career search; these are my favorites.

Create a LinkedIn profile while you are still in the military to help build your network. The veteran community is very active on LinkedIn and it is one of the biggest tools I use when trying to learn or connect with others. Your LinkedIn network can help you find your next career when you are ready. If you work in a career field with a security clearance and want to continue in that career field, make sure to create a ClearanceJobs profile. ClearanceJobs is similar to LinkedIn but is more secure and helps companies fill roles that require security clearances.

THE MILITARY CHANGED MY LIFE

Joining the military was the best choice I ever made for my life. I went from being lost with no path forward to finding purpose and a passion to help others. Military service led me to where I am today and it is how I met my husband. I have had the opportunity to see the world by living across the country, traveling for various trainings, and deploying to Afghanistan. It opened my eyes to a world I did not know existed. Growing up I thought I would live in my hometown my whole life and had no ambition for the future. But the military changed everything; it opened doors to my future I never imagined opening let alone walking through. I can't imagine what my life would have been like had I not decided to join. I have changed and grown so much.

I hope this book helps you in your journey to military service. And I can't wait for you to begin your path to the military to see how it will change your life.

ACKNOWLEDGMENTS

A GIRL'S GUIDE TO MILITARY SERVICE has been a book I have wanted and tried to write a handful of times. It first started out as a short eight-page guide for young women considering military service. When I first created it, my biggest question was not if the resource was needed but how I would find the young women who were considering the military. But when they continued to download it and reach out to me with more questions, I realized there was a lot more to say on this topic than those eight pages.

It took years of writing just to get to the point where I was ready to send off my book proposal. And while I thought it was close to the book I was going to publish, I really had a long way to go. It would not have been possible without the support of the Elva Resa team and specifically Karen Pavlicin, who took time out of her limited schedule to help me zero in on my focus and find the young woman I needed to write to without getting caught up in all the technicalities of the military.

This book also would not have been possible without all the women who believed in me and shared their story on the *Women of the Military* podcast. The stories of women were what drove me to start focusing not only on the transition out of the military but how to help young women as they enter military service. Hearing your stories helped me realize how much I did not know when I joined the military and how many other women were in my shoes when they first entered the military.

I also want to thank my husband. He has always supported me in each new adventure I have taken after leaving the military. He encouraged me to keep pushing forward and never let me give up on the days I was ready to quit. Instead, he offered a listening ear and let me sort through all the ups and downs.

Thanks to my kids, who love telling people their mom served in the military. I love that I have the opportunity to raise strong boys who know women can lead and change the world.

Thank you to my parents, who always wanted the best for me. They pushed me to do something more than the path I was rambling down. I always wonder if the reason I joined the military was because of how quickly my dad drove me to the recruiting station when I mentioned the idea. You believed in me before I believed in myself and I am forever grateful.

And thanks to all the girls and young women who read this book. It may inspire you to military service or it may lead you to another path. Wherever life takes you, know that your choice in considering military service shows your heart of service, not only to your country but to others. Rely on that guiding principle wherever your life takes you next.